Communications and Control Engineering

Springer
London
Berlin
Heidelberg
New York
Barcelona
Hong Kong
Milan
Paris
Santa Clara
Singapore
Tokyo

Published titles include:

Sliding Modes in Control Optimization
V.I. Utkin

Fundamentals of Robotics
M. Vukobratović

Parametrizations in Control, Estimation and Filtering Problems: Accuracy Aspects
M. Gevers and G. Li

Parallel Algorithms for Optimal Control of Large Scale Linear Systems
Zoran Gajić and Xuemin Shen

Loop Transfer Recovery: Analysis and Design
A. Saberi, B.M. Chen and P. Sannuti

Markov Chains and Stochastic Stability
S.P. Meyn and R.L. Tweedie

Robust Control: Systems with Uncertain Physical Parameters
J. Ackermann in co-operation with A. Bartlett, D. Kaesbauer, W. Sienel and R. Steinhauser

Optimization and Dynamical Systems
U. Helmke and J.B. Moore

Optimal Sampled-Data Control Systems
Tongwen Chen and Bruce Francis

Nonlinear Control Systems (3rd edition)
Alberto Isidori

Theory of Robot Control
C. Canudas de Wit, B. Siciliano and G. Bastin (Eds)

Fundamental Limitations in Filtering and Control
María M. Seron, Julio Braslavsky and Graham C. Goodwin

Constructive Nonlinear Control
R. Sepulchre, M. Janković and P.V. Kokotović

A Theory of Learning and Generalization
M. Vidyasagar

Adaptive Control
I.D. Landau, R. Lozano and M.M'Saad

Stabilization of Nonlinear Uncertain Systems
Miroslav Krstić and Hua Deng

Passivity-based Control of Euler-Lagrange Systems
Romeo Ortega, Antonio Loría, Per Johan Nicklasson and Hebertt Sira-Ramírez

Vincent D. Blondel, Eduardo D. Sontag,
M. Vidyasagar and Jan C. Willems (Eds)

Open Problems in Mathematical Systems and Control Theory

Springer

Vincent Blondel, PhD
Institute of Mathematics, University of Leige, Sart Tilman B37, B-4000 Liege, Belgium

Eduardo Sontag, PhD
Department of Mathematics, Rutgers University, Hill Center 724, New Brunswick, NJ 08903, USA

Mathukumalli, Vidyasagar, PhD
Centre for Artificial Intelligence and Robotics, Raj Bhavan Circle - High Grounds, Bangalore 560001, India

Jan C. Willems, PhD
Institute of Mathematics, Rijksuniversiteit te Groningen, Postbus 800, 9700 Av Groningen, The Netherlands

ISBN-13: 978-1-4471-1207-5

British Library Cataloguing in Publication Data
A catalogue record for this book is available from the British Library

Library of Congress Cataloging-in-Publication Data
Open problems in mathematical systems and control theory / V. Blondel ... [et al.].
p. cm. -- (Communications and control engineering)
Includes bibliographical references and index.
ISBN-13: 978-1-4471-1207-5 e-ISBN-13: 978-1-4471-0807-8
DOI: 10.1007/978-1-4471-0807-8
1. Systems analysis 2. Control theory. I. Blondel, Vincent.
II. Series.
QA402.062 1998 98-38067
003--dc21 CIP

Softcover repint of the hardcover 1st edition 1999

Typesetting: Camera ready by editors

69/3830-543210 Printed on acid-free paper

Preface

The idea of inviting the system theory community to formulate some of the open problems in this area of research originated with the first editor. A large number of the problems that appear in this book were first presented in a workshop that was organized in Liège (Belgium) in June 1997. Subsequently, the style of presentation was made somewhat uniform, and together with other contributions, they were refereed. The final result is the book that lies in front of you.

"Open problems" enjoy a great appeal in the mathematical sciences. Certainly the most famous open problem in mathematics is (but one should really say: was) Fermat's last theorem, stating that there is no solution to the equation $a^n + b^n = c^n$, with a, b, c, n positive integers and $n > 2$. This problem remained unsolved for more than three centuries, until Wiles proved the result less than a decade ago.

The most celebrated series of open problems is Hilbert's address to the International Congress of Mathematics at Paris in 1900. In his lecture entitled "Mathematical Problems", Hilbert formulated 23 problems which he felt were the most striking questions in mathematics at that time. These problems ranged from logic and geometry to the calculus of variations and the axiomatization of the laws of physics. In 1976, the American Mathematical Society published a two-volume book on the mathematical developments that arose from Hilbert's problems, a tribute to the enormous mathematical activity that Hilbert's lecture generated. There have been several attempts to, so to say, update Hilbert's list. In the AMS volume mentioned above, 27+ mathematicians formulated 27 problem areas under the title "Problems of Present Day Mathematics". The most recent contribution in this open-problem-vein is the article by Steve Smale in the Spring 1998 issue of *The Mathematical Intelligencer* in which he lists 18 problems under the title "Mathematical Problems for the Next Century".

Open problems are to some extent a prerogative of mature mathematical fields. A 'good' open problem, it is felt, is one that can be stated precisely and briefly, and that requires little background to state but that is very hard to solve. Obviously, Fermat's last theorem and other problems from number theory are jewels in this respect. Continuous mathematics will usually depend on more background notions, and when it comes to important problems of an undeniable mathematical nature as the formulation of the foundations of statistics or of quantum mechanics, the road becomes more slippery. Indeed, who will ever decide that the question has been answered correctly?

It is against the background described in the above two paragraphs that this book should be read. Many of the problems that are put forward bear some of the features of good open problems, although in system theory the questions usually involve continuous mathematics and as such require a number of specialized notions. We believe that the problems are challenging and important. For background material on the subjects treated in this book, the reader may wish to consult some of the many textbooks dealing with the mathematical foundations of the field. Three random references which come to mind are:

J.W. Polderman and J.C. Willems, Introduction to Mathematical Systems Theory: A Behavioral Approach, Springer, NY, 1998.

Eduardo D. Sontag, Mathematical Control Theory: Deterministic Finite Dimensional Systems, Second Edition, Springer, NY, 1998.

M. Vidyasagar, A Theory of Learning and Generalization, Springer, London, 1997.

Three mirror web sites have been set up for posting follow-up information, comments, and solutions:

http://www.ulg.ac.be/mathsys/blondel/openprobs.html

http://www.math.rutgers.edu/~sontag/openprobs.html

http://www.cair.res.in/~sagar/openprobs.html

As we already stated, good open problems are a prerogative of mature, mathematical fields. System theory, however, is only 40 years old and, as such, it can hardly be called mature. It is also only in part mathematical: it relies heavily on problems coming from applications, especially applications to control, and, as such, it is technology-driven. It is therefore entirely acceptable to treat these open problems in the way problems in this field have been treated in the past. After reformulation, simplification and modification, the open problem leads to a solution of a different problem that is perhaps easier but that is perhaps also more important.

We wish the reader much enjoyment and stimulation in reading this book.

The editors,
Liège, New Brunswick, Bangalore, and Groningen, June 1998.

Contents

1

Uniform asymptotic stability of linear time-varying systems

Dirk Aeyels and Joan Peuteman

Universiteit Gent
SYSTeMS
Technologiepark-Zwijnaarde 9
9052 Gent
BELGIUM
dirk.aeyels@rug.ac.be
joan.peuteman@rug.ac.be

1 Description of the problem

Let $\dot{x}(t) = A(t)x(t)$, with $x : \mathbf{R} \to \mathbf{R}^n$ and $A : \mathbf{R} \to \mathbf{R}^{n \times n}$, be a linear time-varying system. We consider the case where the eigenvalues of the system matrix as functions of the time $t \in \mathbf{R}$ may have positive real parts. Consider the following examples: the matrix $A_1(t) := \mathrm{diag}(-1 + 2\cos t, -1 - 2\cos t)$ and the matrix $A_2(t) := \mathrm{diag}(-1 + \max\{2\cos t, -2\cos t\}, -1 - \max\{2\cos t, -2\cos t\})$. Notice that $A_1(t)$ and $A_2(t)$ have the same eigenvalues $\forall t \in \mathbf{R}$. The system $\dot{x}(t) = A_1(t)x(t)$ is uniformly asymptotically stable and the system $\dot{x}(t) = A_2(t)x(t)$ is unstable. This illustrates that the stability properties do not only depend on the evolution of the eigenvalues. The evolution of the corresponding eigenvectors will also play a crucial role.

We are interested in formulating a sufficient condition (preferably easily verifiable) for asymptotic stability of $\dot{x}(t) = A(t)x(t)$. Since this problem turns out to be non-trivial and since specific versions of this problem are currently being examined within the framework of complexity theory [9], we will restrict our attention to a particular class $\dot{x}(t) = -f(t)m(t)m^T(t)x(t)$ with $m : \mathbf{R} \to \mathbf{R}^n$ and $f : \mathbf{R} \to \mathbf{R}$ taking both negative and positive values. Even this problem turns out to be non-trivial.

2 Available results and desired extensions

1. A fast time-varying system $\dot{x}(t) = A(\alpha t)x(t)$ with α sufficiently large and with the additional properties that $A(t)$ and its derivatives up to the second order are continuous and bounded, is exponentially stable when the averaged system $\dot{x}(t) = \bar{A}x(t)$ is exponentially stable [2, 4, 10]. When the system matrix $A(t)$ is periodic with period T, then $\bar{A} = \frac{1}{T}\int_0^T A(t)dt$.
   When the system matrix $A(t)$ is periodic with period T, then the slowly time-varying system $\dot{x}(t) = A(\frac{t}{\alpha})x(t)$ with α sufficiently large is exponentially stable when the average of the real part of the instantaneous eigenvalue $\lambda_{\max}(t)$ is negative i.e. $\frac{1}{T}\int_0^T \lambda_{\max}(t)dt < 0$ [11]. Here, $\lambda_{\max}(t)$ denotes the instantaneous eigenvalue of $A(t)$ with the largest real part.
   The study of the stability of a system $\dot{x}(t) = A(t)x(t)$ with a system matrix $A(t)$ which is neither fast time-varying nor slowly time-varying is an ambitious task. There are some results available in the literature [16, 6] which will be discussed. New contributions in the field are welcome.

2. Wazewski has shown that when $\phi(t, t_0, x_0)$ denotes the solution of $\dot{x}(t) = A(t)x(t)$ with initial condition x_0 at t_0 that $\|\phi(t, t_0, x_0)\| \leq \|x_0\|e^{r(t,t_0)}$ with $r(t, t_0) := \frac{1}{2}\int_{t_0}^t \lambda_{max}(\sigma)d\sigma$. Here $\lambda_{max}(\sigma)$ is the largest eigenvalue of $A(\sigma)+A^T(\sigma)$. For a proof based on Liapunov theory, we refer the reader to [16]. Notice that the system $\dot{x} = A(t)x(t)$ with $A(t) := \text{diag}(-1 + 2\cos t, -1 - 2\cos t)$ is exponentially stable without satisfying the condition of Wazewski. This example illustrates that it is a sufficient condition which is also for systems with a symmetric system matrix, rather conservative. Therefore, it would be useful to obtain a less conservative condition e.g. by an explicit use of the evolution of the eigenvectors.

3. Consider a *class* of n-dimensional systems, each system matrix in companion form with coefficients $p_i(t)$ and with the additional property that $\forall t \in \mathbf{R} : 0 \leq a_i \leq p_i(t) \leq b_i$ $(i = 1, ..., n)$. A necessary and sufficient condition that all solutions of an *arbitrary* equation of this class of systems is nonoscillatory (every nontrivial solution has only a finite number of zeros on $[t_0, \infty[$) and tends to zero as $t \to \infty$ is that the polynomials $u^n + a_1u^{n-1} + b_2u^{n-2} + a_3u^{n-3} + ...$ and $u^n + b_1u^{n-1} + a_2u^{n-2} + b_3u^{n-3} + ...$ have only real roots. These roots are negative since the coefficients of the polynomials are nonnegative. Levin [6] shows that this statement implies that the instantaneous eigenvalues of each system matrix are negative and real. The requirement that the roots are negative is a necessary condition for asymptotic stability of each system belonging to the class (without

this condition one can take an unstable time-invariant system belonging to the class).
It would be useful to obtain a similar stability condition, still implying convergence to zero as $t \to \infty$, but this time for allowing instantaneous complex eigenvalues. Also the question arises whether it is possible to derive similar conditions for other classes of system matrices with coefficients varying between a lower and an upper bound. We believe that these problems are highly non-trivial.

Notice that in the cases 1 and 2, stability conditions for one single system are discussed. In the case 3, the stability condition is formulated for a class of systems.

3 Partial solutions of the open problems

1. Consider the n-dimensional system $\dot{x}(t) = -m(t)m^T(t)x(t)$ with a bounded $m : \mathbf{R} \to \mathbf{R}^n$. When $\exists T > 0$, a $c_1 > 0$ and a $c_2 > 0$ such that $\forall t \in \mathbf{R}$:
$$c_1 I \leq \int_t^{t+T} m(\sigma)m^T(\sigma)d\sigma \leq c_2 I$$
then the system is exponentially stable [3].

2. We would like to derive a sufficient condition for exponential stability of the system $\dot{x}(t) = -f(t)m(t)m^T(t)x(t)$ with a bounded $f : \mathbf{R} \to \mathbf{R}$ taking both positive and negative values. Without loss of generality, we may assume $\|m(t)\| = 1$ for all $t \in \mathbf{R}$. The eigenvectors and eigenvalues are known i.e. $m(t)$ is an eigenvector with eigenvalue $-f(t)$ and the other eigenvectors correspond to the eigenvalue 0 and constitute with $m(t)$ an orthonormal base of $\mathbf{R}^n$. In case $f : \mathbf{R} \to \mathbf{R}$ takes positive and negative values, $\frac{\partial(x^T x)}{\partial x}(-f(t)m(t)m^T(t)x) = -2f(t)(m^T(t)x)^2$ is not always negative. Therefore, $x^T x$ does not qualify as a Liapunov function for proving exponential stability. By the main theorems of [1, 2], a sufficient condition for exponential stability is the existence of a $T > 0$ and a $c > 0$ such that $\forall t \in \mathbf{R}$, $\forall x \in \mathbf{R}^n$: -2 $\int_t^{t+T} f(\sigma)(m^T(\sigma)\phi(\sigma, t, x))^2 d\sigma \leq -c\|x\|^2$. This condition is hard to verify since one needs the value $\phi(\sigma, t, x)$ at time σ of the trajectory corresponding to initial state x at t.

3. We restrict our attention to the two-dimensional case and denote the normed vector perpendicular to $m(t)$ as $p(t)$. Without loss of generality, define $\theta(t)$ such that $m(t) = [\cos(\theta(t)), \sin(\theta(t))]^T$ and therefore $p(t) = [-\sin(\theta(t)), \cos(\theta(t))]^T$. By the time-variant state transformation $z(t) = [p(t)\ m(t)]x(t)$, one obtains that
$$\dot{x}(t) = -f(t)m(t)m^T(t)x(t),$$

with $x : \mathbf{R} \to \mathbf{R}^2$ is equivalent with

$$\dot{z}(t) = \begin{pmatrix} 0 & -\dot{\theta}(t) \\ \dot{\theta}(t) & -f(t) \end{pmatrix} z(t)$$

Since $\|x(t)\| = \|z(t)\|$, one representation is exponentially stable if and only if the other representation is exponentially stable. When for all $t \in \mathbf{R}$: $\dot{\theta}(t) = -1$, the transformed system is a model for the small oscillations of the pendulum in presence of friction $f(t)$. When $f(t)$ is replaced by $f(\frac{t}{\alpha})$ with α large then Solo [11] provides a sufficient condition for exponential stability. When $f(t)$ is replaced by $f(\alpha t)$ with α large, then by averaging theory [2, 10] it follows immediately that the z-system is exponentially stable if $\bar{f}$ (the average of f) is positive. In case f is neither fast time-varying nor slowly time-varying and takes positive and negative values, the stability analysis is non-trivial.

4. In an attempt to extend the result under 3, we consider the z-system where we leave $\dot{\theta}(t)$ to be a bounded function of t but replace $f(t)$ by $f(\alpha t)$ with α large. (This corresponds in x-coordinates to $\dot{x}(t) = -f(\alpha t)m(t)m^T(t)x(t)$) By the main theorems of [1, 2], it is possible to prove that this z-system is exponentially stable if the z-system with $f(t)$ replaced by $\bar{f}$ is exponentially stable. Because of the equivalence with the representation in x-coordinates, this is true if $\bar{f} > 0$ and $\exists T > 0$, a $c_1 > 0$ and a $c_2 > 0$ such that for all $t \in \mathbf{R}$: $c_1 I \leq \int_t^{t+T} m(\sigma)m^T(\sigma)d\sigma \leq c_2 I$ [3]. The property that exponential stability of the z-system with $f(t)$ replaced by $f(\alpha t)$ (with α large) is implied by exponential stability of the z-system with $f(t)$ replaced by $\bar{f}$ is an extension of the averaging results of [2, 4, 10] and will be further treated in [7].

4 References

[1] D. Aeyels and J. Peuteman, "A new asymptotic stability criterion for non-linear time-variant differential equations", *to appear in IEEE Trans. Automat. Control.*

[2] D. Aeyels and J. Peuteman, "On exponential stability of non-linear time-variant differential equations", *submitted for publication.*

[3] B.D.O. Anderson, "Exponential stability of linear equations arising in adaptive identification", *IEEE Trans. Automat. Control*, 22, pp.84-88 (1977).

[4] R.W. Brockett, "Finite dimensional linear systems", John Wiley and sons inc., pp.206-207 (1970).

[5] H.K. Khalil, "Nonlinear systems", Printice Hall, pp.345-346 (1996).

[6] A.Y. Levin, "Absolute nonoscillatory stability and related questions", *St. Petersburg Math. J.*, vol. 4 no. 1. pp.149-161 (1993).

[7] J. Peuteman and D. Aeyels, "Exponential stability of non-linear time-variant differential equations and partial averaging", *submitted for publication.*

[8] V. Solo, "On the stability of Slowly Time-Varying Linear Systems", *Mathematics of Control, Signals, and Systems*, 7 , pp. 331-350 (1994).

[9] J.N. Tsitsiklis and V.D. Blondel, "The Lyapunov exponent and joint spectral radius of pairs of matrices are hard —when not impossible— to compute and to approximate", *Mathematics of Control, Signals, and Systems*, 10, pp. 31-40 (1997).

[10] J.L. Willems, "Stability theory of dynamical systems", Nelson, pp. 117-119 (1970).

2

Positive system realizations

Brian D. O. Anderson[1]

Department of Systems Engineering
Research School of Information Sciences and Engineering
Australian National University
Canberra, ACT 0200
AUSTRALIA
brian.anderson@anu.edu.au

1 Description of the problem

Consider a rational discrete-time transfer function $H(z)$ with the property that the associated causal impulse response $h(k)$ is nonnegative for all k. A positive realization is a quadruple A, b, c, d such that

$$H(z) = d + c(zI - A)^{-1}b$$

and all entries of A, b, c, d are nonnegative.
Easily checkable necessary and sufficient conditions for the existence of such A, b, c, d are known, see [1, 2, 3]. The conditions require checking of the poles of $H(z)$, and a sufficient condition is that $H(z)$ have a single pole of maximum modulus. There are rational $H(z)$ with nonnegative $h(k)$ which have no positive realization, but they are in a sense nongeneric. A number of open problems relating to realization however remain:

1. How may one straightforwardly determine the least dimension of any positive realization? In particular, when is this dimension equal to the McMillan degree of $H(z)$?

2. How are positive realizations of least dimension related? In particular, apart from reordering of the states, is it true or not that the set of least dimension positive realizations is a connected set?

[1]The author wishes to acknowledge the funding of the activities of the Cooperative Research Centre for Robust and Adaptive Systems by the Australian Commonwealth Government under the Cooperative Research Centres Program.

3. For an $n-th$ order $H(z)$ for which the least dimension of positive realizations is N, what is the maximum number (generically) of zeros entries that can be included in $\{A, b, c\}$? In particular, can canonical forms for positive realizations be found?

4. How can one approximate a high order positive realization by a low order positive realization where the measure of approximation reflects either transfer function approximation error on the unit circle, or impulse response approximation error.

2 History of the problem

The problem of positive realization has been around for some time; significant progress towards a solution was made in [4], where the existence of a positive realization was cast in terms of the existence of a polyhedral cone with certain properties. Knowledge of a cone almost immediately yields a solution of the construction problem.
References [1, 2, 3], gives conditions on the poles of $H(z)$ for the existence of the cone. Two examples in the survey [5] are:

$$\begin{aligned} H(z) &= -\frac{1}{z+0.8} + \frac{1}{z-0.8} + \frac{1}{z+0.5} \\ h(2k) &= 2[(0.8)^{2k-1}] - (0.5)^{2k-1} \qquad k = 1, 2 \ldots \\ h(2k+1) &= (0.5)^{2k} \qquad k = 1, 2 \ldots \end{aligned}$$

No 3rd order positive realization exists, but a 4th order one does. Second, for the transfer function/impulse response pair

$$H(z) = \frac{1}{2} = \left[\frac{\frac{1}{2}}{z - \frac{1}{2}} - \frac{z(\frac{1}{2}\cos 2) - \frac{1}{4}}{z^2 - z\cos 2 + \frac{1}{4}} \right]$$

$$h(k) = \left(\frac{1}{2}\right)^k sin^2 k$$

no (finite dimensional) positive realization exists.
Reference [6] has the example

$$H(z) = \frac{1}{z-1} - 25\frac{(0.4)^{4-N}}{z-0.4} + 75\frac{0.2^{4-N}}{z-0.2}$$

for which the least dimension of a positive realization is N, assuming $N \geq 4$. These examples perhaps give insight into the subtle nature of the problem.

3 Motivations

Questions 1 and 2 are clearly of a basic systems theory nature, and are of interest in their own right. Additionally, there is a technology termed "charge-routing networks" [7] which is available to realize a positive realization of a digital filter transfer function. Questions 1 and 2 are relevant in selecting a way to realize a positive realization. Lest it be thought that the technology is of too limited applicability (as digital transfer functions are perhaps unlikely to have nonnegative impulse responses), we comment that any transfer function $H(z)$ can be expressed as

$$H(z) = H_1(z) - H_2(z)$$

where H_1 and H_2 have nonnegative impulse responses, H_1 has degree one more than H and H_2 has degree 1.
Questions 2 and 3 are relevant in considering problems of identification, especially of compartmental systems, which are associated with positive realizations, [8].
Question 4 is motivated by a related but even harder problem of great practical significance: how may one approximate a hidden Markov model having a large number of states by a hidden Markov model with a small number of states?
The connection between hidden Markov models and positive systems is explained in, for example, [5].

4 References

[1] B.D.O. Anderson, M. Deistler, L. Farina, and L. Benevuti, "Nonnegative realization of a linear system with nonnegative impulse response", *IEEE Transactions on Circuits and Systems - I: Fundamental Theory and Applications*, Vol 43, 1996, pp 134-142.

[2] T. Kitano and H. Malda, "Positive realization by down sampling the impulse response", preprint 1996.

[3] L. Farina, "On the existence of a positive realisation", *Systems and Control Letters*, Vol 28, 1996, pp 219-226.

[4] Y. Ohta, H. Maeda, and S. Kodama, "Reachability, observability and realizability of continuous-time positive systems", *SIAM Journal Control Optimization*, Vol 22, 1984, pp 171-180.

[5] B.D.O. Anderson, "New developments in the theory of positive systems", in *Systems and Control in the Twenty-First Century*, C. Byrnes, B. Datta, M. Gilliam, and C. Martin. eds., Birkhauser, Boston, 1997.

[6] L. Benvenuti, and L. Farina, "Examples of positive realizations of minimal order having an arbitrarily large dimension", submitted for publication.

[7] A. Gersho, B. Gopinath, "Charge-routing networks", *IEEE Transactions, Circuits and Systems*, Vol CAS-26, 1979, pp 81-92

[8] S. Rinaldi, and L. Farina, "Positive Linear Systems: theory and applications", (in Italian), Citta Studi, Milan, 1995.

3

System approximation in the 2-norm

A.C. Antoulas

Department of Electrical Engineering
Rice University
Houston, TX 77251
USA
aca@rice.edu

1 Finite-dimensional operators

We present below a summary of some open problems in systems and control related to system approximation. For more details we refer to [3].

Given $A \in \mathbb{R}^{n\times n}$, rank $A = r \leq n$, find $X \in \mathbb{R}^{n\times n}$, rank $X = k < r$, such that the induced 2-norm of the *error* matrix $E := A - X$ is minimized. The solution of this non-convex problem, known as the optimal approximation in the induced 2-norm, is given by the *Schmidt-Mirsky* theorem; in involves the *singular value decomposition* of A:

$$\min_{\text{rank}X \leq k} \| A - X \|_2 = \sigma_{k+1}(A) \tag{3.1}$$

where $\sigma_i(A) \geq \sigma_{i+1}(A)$, $i = 1, \cdots, n-1$, are the ordered *singular values* of A. One minimizer X_*, which is non-unique, is obtained by appropriate truncation of the dyadic decomposition of A.
The *sub-optimal approximation* problem in the 2-induced norm is, given ϵ, to find X satisfying

$$\sigma_{k+1}(A) < \| A - X \|_2 < \epsilon < \sigma_k(A) \tag{3.2}$$

The solution to this problem is due to [6]. We present here its essential features. X satisfies (3.2) if, and only if, there exists a J-unitary matrix Θ of size $2n$, and a contraction Δ such that

$$A - X = E := E_1 E_2^{-1}, \quad \| \Delta \|_2 < 1 \quad \text{where} \tag{3.3}$$

$$\begin{pmatrix} E_1 \\ E_2 \end{pmatrix} := \Theta \begin{pmatrix} \Delta \\ I_n \end{pmatrix} = \begin{pmatrix} \Theta_{11}\Delta + \Theta_{12} \\ \Theta_{21}\Delta + \Theta_{22} \end{pmatrix} \tag{3.4}$$

The J-unitary matrix Θ can be constructed using ϵ and the singular value decomposition of A.

2 Infinite-dimensional Hankel operators

Consider now *Hankel operators* defined as follows:

$$\mathcal{H}:\ \ell_2(\mathbb{Z}_+) \longrightarrow \ell_2(\mathbb{Z}_+) \ \text{ where } \ (\mathcal{H})_{ij} := \alpha_{i+j-1} \in \mathbb{R},\ i,j>0 \tag{3.5}$$

for some sequence α_t, $t = 1, 2, \cdots$. We will assume that

$$\operatorname{rank}\mathcal{H} = n < \infty \ \text{ and } \ \sum_{t>0} |\ \alpha_t\ |^2 < \infty$$

Realization theory of linear time-invariant systems establishes the equivalence of sequences whose associated Hankel matrix has finite rank and linear time-invariant, finite dimensional systems expressed in terms of their transfer functions G or in terms of the triple (C, A, B):

$$\mathcal{H},\ \operatorname{rank}\mathcal{H} = n \ \overset{1-1}{\longleftrightarrow}\ G(z) = \frac{\pi(z)}{\mathcal{H}i(z)} = C(zI - A)^{-1}B,\ \ \deg\pi < \deg\chi = n \tag{3.6}$$

The square summability of the sequence α_t, implies that the roots of χ are inside the unit disc.
The problem is to approximate $\mathcal{H}$ by a Hankel matrix $\hat{\mathcal{H}}$ of lower rank; optimality is sought in the 2-induced norm of the Hankel operator $\mathcal{H}$. By the equivalence (3.6), this corresponds to approximating the linear system described by $G = \frac{\pi}{\mathcal{H}i}$, by a system of lower complexity, i.e. $\hat{G} = \frac{\hat{\pi}}{\hat{\mathcal{H}i}}$, where $\deg\mathcal{H}i > \deg\hat{\mathcal{H}i}$. This problem is known as *linear system approximation in the Hankel norm.* For details see e.g. [5]. Let the ordered non-zero singular values of $\mathcal{H}$ be $\sigma_i(\mathcal{H}) \geq \sigma_{i+1}(\mathcal{H})$, $i = 1, \cdots, n-1$. The question arises: *does there exist an approximant of rank k which has Hankel structure and achieves the lower bound $\sigma_{k+1}(\mathcal{H})$ predicted by the Schmidt-Mirsky result?* The answer is given by the AAK (Adamjan, Arov, Krein) theory [1], [2]. It asserts the existence of a unique approximant $\hat{\mathcal{H}}$ of rank k, which has Hankel structure and attains the lower bound, namely

$$\sigma_1(\mathcal{H} - \hat{\mathcal{H}}) = \sigma_{k+1}(\mathcal{H})$$

As in the finite-dimesional operator case the problem of suboptimal solutions arises, namely, given ϵ, find all $\hat{\mathcal{H}}$ having Hankel structure and rank k, such that

$$\sigma_{k+1}(\mathcal{H}) \leq \sigma_1(\hat{\mathcal{H}} - \mathcal{H}) < \epsilon < \sigma_k(\mathcal{H}) \tag{3.7}$$

The solution to (3.7) (see pages 540-546 of [4], and [5]) is given in terms of the realization (C, A, B) of the rational (transfer) function G attached to the Hankel matrix $\mathcal{H}$ by means of (3.6).

Recall that the AAK theory is valid for Hankel operators $\mathcal{H}$ whose associated rational function G is *stable*. The solution to (3.7) however makes use of rational functions which have poles *outside* the unit disc, i.e. are *anti-stable*. A rational function F with no poles on the unit circle, can be decomposed uniquely as a sum of an anti-stable and a stable rational function: $F = [F]_- + [F]_+$, where $[F]_-$ has all its poles outside the unit disc and $[F]_+$ has all its poles inside the unit disc and is strictly proper.
Let E_i, $i = 1, 2$, and Δ be related by:

$$E_1(\Delta)E_2(\Delta) := \Theta\Delta I = \begin{pmatrix} \Theta_{11}(z)\Delta(z) + \Theta_{12}(z) \\ \Theta_{21}(z)\Delta(z) + \Theta_{22}(z) \end{pmatrix} \tag{3.8}$$

where Θ is a 2×2 matrix which is J-unitary on the unit circle, i.e. $\Theta(e^{j\theta})J\Theta^*(e^{-j\theta}) = J$, and J is the 2×2 signature matrix $J := \text{diag}\,(1, -1)$. The following result holds true: $\hat{\mathcal{H}}$ satisfies (3.7) if, and only if, there exists an anti-stable contraction Δ, ($|\ \Delta(e^{j\theta})\ | < 1$), such that

$$G_{\hat{\mathcal{H}}} = [F]_+ \text{ where } G_{\mathcal{H}} - F = E(\Delta) := E_1(\Delta)E_2^{-1}(\Delta) \tag{3.9}$$

It should be mentioned that the optimal and sub-optimal approximation results above, hold equally for operators with *block Hankel* structure (i.e. $\alpha_t \in \mathbb{R}^{p\times m}$). See [5] and [4] for details.

3 Open problems

We have presented above two cases of optimal and sub-optimal approximation in the 2-induced norm, of linear operators of finite rank, namely that of *unstructured operators in a finite-dimensional space*, and that of *structured (Hankel) operators in an infinite-dimensional space.* In the parametrization of all sub-optimal solutions *two* important ingredients manifest themselves **in both cases**: (a) Linear fractions in terms of an appropriately defined J-unitary matrix Θ, and (b) all solutions are parametrized in terms of an appropriate contraction Δ.
The problem of deriving an optimal or sub-optimal reduced order model of a system from exact (or noisy) input-output measurements is of importance for system identification and robust control. In [5] this problem is studied for the special case where the input to the system is an impulse and the output the impulse response; for the more general case of arbitrary input-output measurements the problem is studied in [7].
From the above considerations, the following problems arise:

1. Investigate the existence of a unifying framework for the approximation of operators in the induced 2-norm. Special cases of this framework should include (3.3), (3.4) as well as (3.8), (3.9).

2. Investigate the problem of low rank approximation of structured finite matrices, and in particular Hankel matrices, in the induced 2-norm. In particular, when is the minimum predicted by the Schmidt-Mirsky result attained, when not, and what does this depend on?

3. Investigate the problem of obtaining reduced order models of a system from arbitrary measurements - finite or infinite - of the input and the output, which are optimal and/or suboptimal in the induced 2-norm.

4 References

[1] V.M. Adamjan, D.Z. Arov, and M.G. Krein, *Analytic properties of Schmidt pairs for a Hankel operator and the generalized Schur-Takagi problem*, Math. USSR Sbornik, **15**: 31-73 (1971).

[2] V.M. Adamjan, D.Z. Arov, and M.G. Krein, *Infinite block Hankel matrices and related extension problems*, American Math. Society Transactions, **111**: 133-156 (1978).

[3] A.C. Antoulas, *Approximation of linear operators in the 2-norm*, Linear Algebra and its Applications, Special Issue on "Challenges in Linear Algebra", to appear (1998).

[4] J.A. Ball, I. Gohberg, and L. Rodman, *Interpolation of rational matrix functions*, Operator Theory: Advances and Applications, Birkhäuser (1990).

[5] K. Glover, *All optimal Hankel-norm approximations of linear multivariable systems and their L^∞-error bounds*, International Journal of Control, **39**: 1115-1193 (1984).

[6] A.J. van der Veen, *A Schur method for low rank matrix approximation*, SIAM J. Matrix Analysis and Applications, January (1996).

[7] J.C. Willems, *From time series to linear system - Part III: Approximate modeling*, Automatica **23**: 87-115 (1987).

4

Is it possible to recognize local controllability in a finite number of differentiations?

Andrei A. Agrachev

Steklov Mathematical Institute
ul. Gubkina 8
Moscow 117966
RUSSIA
agrachev@mi.ras.ru

1 Description of the problem

Let $f_1, \dots, f_k$, $k \geq 2$, be real–analytic vector fields defined on a neighborhood of the origin in R^n, and $t > 0$. The point $x \in R^n$ is called attainable from the origin for a time less than t and with no more than N switchings, if there exists a subdivision $0 = t_0 < t_1 < \ldots < t_{N+1} < t$ of the segment $[0, t]$ and solutions $\xi_j(t)$, $t \in [t_j, t_{j+1}]$ of the differential equations $\dot{x} = f_{i_j}(x)$, for some $i_j \in \{1, \dots, k\}$, such that $\xi_0(0) = 0$, $\xi_{j-1}(t_j) = \xi_j(t_j)$ for $j = 1, \dots, N$, $\xi_N(t_{N+1}) = x$. Let $\mathcal{A}_t(N)$ be the set of all such points x; the set $\mathcal{A}_t = \bigcup_{N>0} \mathcal{A}_t(N)$ is called the attainable set for a time no greater than t.

The family of vector fields $\{f_1, \dots, f_k\}$ is said to be *Small Time Locally Controllable* (STLC) at $0 \in R^n$, if $\mathcal{A}_t$ contains an open neighborhood of 0 for any positive t, i. e. $0 \in int\, \mathcal{A}_t \quad \forall t > 0$.

Let $\{f_1, \dots, f_k\}$ be an STLC family of vector fields. Is it true that:

(a) *$\exists\, N_1$ such that any family of vector fields with the same Taylor polynomials of order N_1 at 0 as the Taylor polynomials of $f_1, \dots, f_k$ is STLC ?*

(b) *$\exists\, N_2$ such that $\mathcal{A}_t$ contains a ball of radius t^{N_2} centered at 0, for any small enough $t > 0$?*

(c) *$\exists\, N_3$ such that $0 \in \mathcal{A}_t(N_3)$, for any $t > 0$?*

The property (c) implies both (a) and (b). The answer to question (c) is positive, if $n \leq 2$, and is negative for some STLC couples $\{f_1, f_2\}$ of polynomial vector fields, if $n \geq 4$. Question (c) is open for $n = 3$. Questions (a) and (b) are open for all $n \geq 3$ for both analytical and polynomial cases.

2 Motivations and references

The local controllability problem is classical in Geometric Control Theory. If one has a controlled system, the first question is: in what direction one can move? Small Time Local Controllability is just a possibility of moving in any direction, slower or faster. Problems (a), (b), (c) have the following meaning:

- Is it always possible to recognize an STLC family in a finite number of differentiations?
- Is it possible to move not too slow in any direction?
- How complicated is the control strategy one really needs?

If $\{f_1, \ldots, f_k\}$ is an STLC family, then $\{-f_1, \ldots, -f_k\}$ is also STLC (this is true though not evident); hence $\{f_1, \ldots, f_k\}$ is small time stabilizable at 0 by open-loop controls. See [15] for the connection with the stabilizability by a feedback control.
There are many partial results. If the cone $\{\sum_{i=1}^{k} \alpha_i f_i \mid \alpha_i \geq 0\}$ is a linear space, then a simple characterization of the STLC families given in [1] provides the positive answers to all the questions (see also [13, 14] for uniform estimates of N_2, N_3 in the case of polynomial vector fields of prescribed degree). The symmetrized family $\{f_1, \ldots, f_k, -f_1, \ldots, -f_k\}$ automatically satisfies the previous condition. If this symmetrized family is STLC, then the attainable sets for the original family $\{f_1, \ldots, f_k\}$ have a nonempty interior, although the family may fail to be STLC ([2]). If $\{f_1, \ldots, f_k\}$ is STLC, then for any $t > 0$ there exists N_t such that $0 \in int\, \mathcal{A}_t(N_t)$ (do not confuse with property (c), where N doesn't depend on t !). The positive answer to all the questions in the 2-dimensional case follow from the results of [3]; a counterexample to property (c) in the 4-dimensional case was obtained in [9].
This is a long story, many efforts were made and rather strong sufficient as well as some necessary conditions for Small Time Local Controllability in a space of arbitrary dimension were obtained, see [4—8], [10—12]. Unfortunately (or maybe fortunately for you, the reader of this chapter), the gap between the necessary and sufficient conditions is still big enough to keep open the above formulated fundamental questions.

3 References

[1] N. Nagano, "Linear differential systems with singularities and applications to transitive Lie algebras", *J. Math. Soc. Japan*, 18, pp. 398–404 (1966).

[2] V. Jurdjevic and H.J. Sussmann, "Controllability of nonlinear systems", *J. Differential Equations*, 12, pp. 95–116 (1972).

[3] N.N. Petrov, "Local controllability", *Differentzial'nye uravnenia*, 12, pp. 2214-2222 (1976).

[4] H. Hermes, "On local controllability", *SIAM J. Control Optim.*, 20, pp. 211–230 (1982).

[5] H. Hermes, "Control systems wich generate decomposable Lie algebras", *J. Differential Equations*, 44, pp. 166–187 (1982).

[6] H.J. Sussmann, "Lie brackets and local controllability: a sufficient condition for scalar input systems", *SIAM J. Control Optim.*, 21, pp. 686–713 (1983).

[7] G. Stefani, "On local controllability of scalar input control systems", in C.I. Byrnes and A. Lindquist (eds), *Theory and Applications of Nonlinear Systems*, North-Holland, Amsterdam, pp. 167–179 (1986).

[8] H.J. Sussmann, "A general theorem on local controllability", *SIAM J. Control Optim.*, 25, pp. 158–194 (1987).

[9] M. Kawski, "Control variations with an increasing number of switchings", *Bull. Amer. Math. Soc.* 18, pp. 149–152 (1988).

[10] M. Kawski, "High-order small time local controllability", in H.J. Sussmann (ed.), *Nonlinear Controllability and Optimal Control*, Marcel Dekker, New York, pp. 431–467 (1990).

[11] A.A. Agrachev and R.V. Gamkrelidze, "Local controllability for families of diffeomorphisms", *Systems Control Lett.* 20, pp. 66–76 (1993).

[12] A.A. Agrachev and R.V. Gamkrelidze, "Local controllability and semigroups of diffeomorphisms", *Acta Appl. Math.* 32, pp. 1–57 (1993).

[13] A. Gabrielov, "Multiplicities of zeroes of polynomials on trajectories of polynomial vector fields and bounds on degree of nonholonomy", *Math. Research Letters* 2, pp. 437–451 (1995).

[14] J.-J. Risler, "A bound on the degree of nonholonomy in the plane", *Theoretical Computer Science* 157, pp. 129–136 (1996).

[15] J.-M. Coron, "Stabilizaton of controllable systems", in A. Bellaiche, J.-J. Risler (eds.), Sub-Riemannian Geometry, Birkhäuser, New York, pp. 365–384 (1996).

5

Open problems in sequential parametric estimation

Er-Wei Bai*, Roberto Tempo** and Yinyu Ye***

*Department of Electrical and Computer Engineering
University of Iowa
Iowa City, Iowa 52242
USA
erwei@icaen.uiowa.edu

**CENS-CNR
Politecnico di Torino
Torino
ITALY
tempo@polito.it

***Department of Management Science
University of Iowa
Iowa City, Iowa 52242
USA
yyye@dollar.biz.uiowa.edu

1 Preliminaries and Problem Statement

In this paper, we consider a single input-single output discrete-time system

$$y_i = \phi_i^T \theta + e_i, \quad i = 1, 2, \ldots, n$$

where $y_i \in \mathbf{R}$ is the system output, $\phi_i \in \mathbf{R}^m$ the measurable regressor, $\theta \in \mathbf{R}^m$ the unknown parameter vector to be identified and $e_i \in \mathbf{R}$ the output noise. This system can be re-written more compactly as

$$y = \Phi\theta + e \tag{5.1}$$

where

$$y = \begin{pmatrix} y_1 \\ y_2 \\ \vdots \\ y_n \end{pmatrix}, \quad \Phi = \begin{pmatrix} \phi_1^T \\ \phi_2^T \\ \vdots \\ \phi_n^T \end{pmatrix} \quad \text{and} \quad e = \begin{pmatrix} e_1 \\ e_2 \\ \vdots \\ e_n \end{pmatrix}.$$

The purpose of parametric system identification is to find an estimate $\hat{\theta}$ of the unknown parameter vector θ from given input-output measurements y and Φ. Most of the research on parametric estimation is focused on the so-called *stochastic* approach where the noise is assumed to be a sequence of random variables with some known probabilistic properties; see e.g. the classical textbook [1]. An alternative method is the *unknown but bounded error* approach, where the a priori information is a given bound $\epsilon > 0$ of the noise, i.e.,

$$|e_i| \leq \epsilon \quad i = 1, 2, \ldots, n. \tag{5.2}$$

Then, the membership-set

$$\Omega = \bigcap_{i=1}^{n} \{\hat{\theta} \in \mathbf{R}^m : -\epsilon \leq y_i - \phi_i^T \hat{\theta} \leq \epsilon\}$$

is the set of all parameter estimates consistent with equation (5.1), the input-output data y and Φ and the noise bound (5.2). For references on this line of research, see for example the special issues [2] and [3]. In the bounded-error context, it is of interest to find specific estimates in the membership-set enjoying certain optimality properties. This is discussed in the next section.

2 Two Optimal Estimates

One of the most popular optimal estimates is the *Chebyshev center* $\hat{\theta}_c$ of the membership-set Ω

$$\hat{\theta}_c = \arg \min_{\theta \in \Omega} \max_{\eta \in \Omega} \|\theta - \eta\|. \tag{5.3}$$

The Chebyshev center is the best worst-case estimate of the true but unknown system parameter vector in the sense that it minimizes the maximum "distance" between $\hat{\theta}_c$ and the unknown parameter vector that generated the data. For given input-output data, the calculation of $\hat{\theta}_c$ is a linear programming problem if the norm in (5.3) is ℓ_∞. It should be pointed out, however, that $\hat{\theta}_c$ is optimal in terms of the parameter estimation error but *not* in terms of the so-called output error $||y - \Phi\hat{\theta}_c||$. To avoid this drawback, the *projection estimate* $\hat{\theta}_p$, also denoted as constrained least squares estimate, is often proposed

$$\hat{\theta}_p = \arg \min_{\hat{\theta} \in \Omega} ||y - \Phi\hat{\theta}||^2. \tag{5.4}$$

If the norm in (5.4) is Euclidean, the projection estimate can be computed with convex programming.

3 Remarks on the Optimal Estimates

1. The Chebyshev center $\hat{\theta}_c$ depends only on the constraints $-\epsilon \leq y_i - \phi_i^T\hat{\theta} \leq \epsilon$ describing the membership-set Ω and *not* on the redundant constraints. The projection estimate $\hat{\theta}_p$ also depends on the constraints describing Ω and on *all* the input-output data y and Φ which enter into the objective function.

2. It is well-known that the unconstrained least squares estimate is in general *not* in the membership-set. Thus, by the convexity of the set Ω and the form of the cost function, we conclude that $\hat{\theta}_p$ is often at the boundary of the membership-set.

3. If the norm in (5.4) is Euclidean, $\hat{\theta}_p$ is the estimate within Ω minimizing the "average" output error

$$\sum_{i=1}^{n}(y_i - \phi_i^T\hat{\theta})^2$$

or, equivalently, maximizing the complementary output error

$$\sum_{i=1}^{n}(\epsilon^2 - (y_i - \phi_i^T\hat{\theta})^2).$$

Thus, the projection estimate $\hat{\theta}_p$ can be immediately re-written as

$$\hat{\theta}_p = \arg\max_{\hat{\theta}\in\Omega}\sum_{i=1}^{n}(\epsilon^2 - (y_i - \phi_i^T\hat{\theta})^2).$$

4 Open Problems in Sequential Parametric Estimation

As previously discussed, the off-line computation of the Chebyshev center $\hat{\theta}_c$ and of the projection estimate $\hat{\theta}_p$ can be easily performed using linear and convex programming, respectively. However, due to the nature of the identification problem, the input-output data is observed *sequentially* in time. In this regard, we observe that the Chebyshev center is a continuous (but not necessarily smooth) function of the data y_i and ϕ_i so that the change in $\hat{\theta}_c$ corresponding to new observed data could not be easily followed. We also notice that $\hat{\theta}_p$ is a continuous function of the data y_i and ϕ_i and in the literature there is no sequential algorithm for computing this estimate. Roughly speaking, by a sequential algorithm, we mean the following: When the new data y_i and ϕ_i is available, the calculation of $\hat{\theta}_p$ at time

i should use the estimate $\hat{\theta}_p$ at the previous time $(i-1)$. More precisely, if the computational cost of a non-sequential algorithm for computing $\hat{\theta}_p$ at time i is $c(i)$, then a sequential algorithm should produce a sequence of estimates $\hat{\theta}_p$ at time $1, 2, \ldots, i$ with a total cost bounded by $O(c(i))$.
The open problem we propose is therefore the computation of these estimates sequentially. That is, letting $\hat{\theta}_c(i)$ and $\hat{\theta}_p(i)$ denote the Chebyshev and the projection estimates at time i, respectively, the goal is to compute them using the estimates $\hat{\theta}_c(i-1)$ and $\hat{\theta}_p(i-1)$ and the current input-output data. Since this problem appears too difficult, we now propose a modification. That is, instead of estimating *exactly* $\hat{\theta}_c(i)$ and $\hat{\theta}_p(i)$, we aim to compute only a δ-approximation. More precisely, letting $\delta > 0$ be a given "arbitrarily small" constant and i_0 be a learning period, the goal is to compute sequentially $\hat{\theta}_c^\delta(i)$ and $\hat{\theta}_p^\delta(i)$ such that

$$||\hat{\theta}_c^\delta(i) - \hat{\theta}_c(i)|| \leq \delta \text{ and } ||\hat{\theta}_p^\delta(i) - \hat{\theta}_p(i)|| \leq \delta$$

for all $i \geq i_0$. In a certain sense, this latter problem is a continuation of the work [4] where a δ-approximation of *interpolatory* algorithms is constructed sequentially. Interpolatory algorithms are estimates within the membership set which are "almost" optimal within a factor of two; see [5]. To conclude this paper, we recall that in [6] a recursive implementation of the so-called analytic center of the membership-set is shown.

5 References

[1] L. Ljung, "System Identification: Theory for the Users," *Prentice-Hall*, Englewood Cliffs, NJ, 1987.

[2] Special Issue on System Identification for Robust Control design, *IEEE Transactions on Automatic Control*, Vol. 37, No. 7, 1992.

[3] Special Issue on Trends in System Identification, *Automatica*, Vol. 31, No. 12, 1995.

[4] E.-W. Bai, K. M. Nagpal and R. Tempo, "Bounded-Error Parameter Estimation: Noise Models and Recursive Algorithms," *Automatica*, Vol. 32, pp. 985-999, 1996.

[5] J. F. Traub, G. Wasilkowski and H. Woźniakowski, "Information-Based Complexity," *Academic Press*, New York, 1988.

[6] E.-W. Bai, Y. Ye and R. Tempo, "Bounded Error Parameter Estimation: A Sequential Analytic Center Approach," *Proceedings of the IEEE Conference on Decision and Control*, San Diego, December 1997.

6
Conditions for the existence and uniqueness of optimal matrix scalings

Venkataramanan Balakrishnan

School of Electrical and Computer Engineering
Purdue University
West Lafayette, IN 47907-1285
USA
ragu@ecn.purdue.edu

1 Description of the problem

Given $M \in \mathbf{C}^{n\times n}$, consider the problem of finding

$$f_{\min}(M) = \inf \left\{ \|DMD^{-1}\| \;\middle|\; D \in \mathcal{D} \right\}, \tag{6.1}$$

where $\|\cdot\|$ denotes the spectral norm, and $\mathcal{D}$ is a set of "scalings" defined by

$$\mathcal{D} = \left\{ D \;\middle|\; \begin{array}{l} D \in \mathbf{C}^{n\times n},\ D = \mathrm{diag}(D_1, \ldots, D_m, d_1 I_{l_1}, \ldots, d_p I_{l_p}) \\ D_i = D_i^* \in \mathbf{C}^{k_i \times k_i},\ d_i \in \mathbf{R},\ \mathrm{Trace}(D) = 1 \end{array} \right\}. \tag{6.2}$$

Let the set of *optimal* scalings $\mathcal{D}_{\mathrm{opt}}$ be defined by

$$\mathcal{D}_{\mathrm{opt}} \triangleq \left\{ D \;\middle|\; D \in \mathcal{D},\ \|DMD^{-1}\| = f_{\min}(M) \right\}. \tag{6.3}$$

We are interested in the following two questions:

1. When is the set $\mathcal{D}_{\mathrm{opt}}$ nonempty?

2. When is the set $\mathcal{D}_{\mathrm{opt}}$ a singleton?

An affirmative answer to the first question means that optimal scalings exist, and an affirmative answer to the second question means that there is a unique optimal scaling.

2 Motivations

Problem (6.1) arises in the robustness analysis of control systems with structured uncertainties. For further details, see references [1, 2]. Problem (6.1) also appears in the context of finding optimal (with various criteria for optimality) preconditioners for use in iterative algorithms; see for example, [3, 4].

3 Available results

For the special case when the set $\mathcal{D}$ consists of diagonal scalings (that is, when $m = 0$, and $l_1 = l_2 = \cdots = l_p = 1$ in (6.2)), partial answers to the questions are available in [5]:

1. $\mathcal{D}_{\text{opt}}$ is nonempty if the matrix M is irreducible, i.e., there does not exist a permutation similarity transformation that renders M block upper-triangular. (A similar result was proved in Proposition 4 of [6] for the more general case of arbitrary l_i, with $m = 0$.)

2. For an irreducible matrix M, let D be an optimal scaling, and let the maximum singular value of DMD^{-1} be simple. Then D is the only optimal scaling if and only if there exist no pair of vectors u and v, with $\|u\| = \|v\| = 1$, satisfying

$$\begin{aligned} DMD^{-1}\,v &= \gamma\,u \\ D^{-1}M^*D\,u &= \gamma\,v, \end{aligned} \tag{6.4}$$

that belong to the same coordinate subspace, i.e., a subspace of the form $\bigcup_{i \in \mathbf{I}} \text{span}\{e_i\}$, where $\mathbf{I}$ is a proper subset of the set of indices $\{1, \ldots, n\}$ and $\{e_i,\ i = 1, \ldots, n\}$ are coordinate vectors, that is, Euclidean unit vectors in $\mathbf{R}^n$.

In other words, *sufficient* conditions for the existence and uniqueness of optimal diagonal matrix scalings are given in [5]; it is also shown there that these conditions are *not* necessary.

4 References

[1] M. G. Safonov, "Stability margins of diagonally perturbed multivariable feedback systems," *IEE Proc.*, vol. 129-D, pp. 251–256 (1982).

[2] J. Doyle, "Analysis of feedback systems with structured uncertainties," *IEE Proc.*, vol. 129-D, pp. 242–250 (1982).

[3] G. E. Forsythe and E. G. Straus, "On best conditioned matrices," *Proc. Amer. Math. Soc.*, vol. 6, pp. 340–345 (1955).

[4] A. Greenbaum and G. H. Rodrigue, "Optimal preconditioners of a given sparsity pattern," *Bit*, vol. 29, pp. 610–634 (1989).

[5] V. Balakrishnan and S. Boyd, "Existence and uniqueness of optimal matrix scalings," *SIAM J. on Matrix Analysis and Applications*, vol. 16, pp. 29–39 (1995).

[6] M. K. H. Fan and A. L. Tits, "m-form numerical range and the computation of the structured singular value," *IEEE Trans. Aut. Control*, vol. AC-33, pp. 284–289 (1988).

7

Matrix inequality conditions for canonical factorization of rational transfer function matrices

Venkataramanan Balakrishnan

School of Electrical and Computer Engineering
Purdue University
West Lafayette, IN 47907-1285
USA
ragu@ecn.purdue.edu

1 Description of the problem

A *canonical factorization* of a square real-rational transfer function matrix $M(s)$ is the following:

$$M(s) = M_+(s)M_-(s),$$

where $M_+(s)$, $M_+(s)^{-1}$, $M_-(-s)$, and $M_-(-s)^{-1}$ belong to $\mathbf{RH}_\infty$.
A related, perhaps more familiar, factorization is *spectral factorization* (see for example, [1, 2]): For a real-rational transfer function matrix $N(s)$ that satisfies $N(s) = N(-s)^T$, a spectral factorization is

$$N(s) = N_+(-s)^T N_+(s),$$

where $N_+(s)$ and $N_+(s)^{-1}$ belong to $\mathbf{RH}_\infty$.
Given a real-rational transfer function matrix M, it is well-known that a necessary and sufficient condition for the existence of a spectral factorization of $M(s) + M(-s)^T$ is that for some $\epsilon > 0$,

$$M(j\omega) + M(j\omega)^* \geq 2\epsilon I \text{ for all } \omega \in \mathbf{R}. \tag{7.1}$$

Given any minimal state-space realization (A, B, C, D) of M, this condition is equivalent to the existence of a real symmetric matrix P such that the following linear matrix inequality holds for some $\epsilon > 0$ (see for example, [2, 3, 4]):

$$\begin{bmatrix} A^TP + PA & PB - C^T \\ B^TP - C & 2\epsilon I - (D + D^T) \end{bmatrix} \leq 0. \tag{7.2}$$

(This is just the Kalman-Yakubovich-Popov-Anderson Lemma.)
It turns out that if condition (7.1) holds for some $\epsilon > 0$, then a *canonical* factorization exists for M. (This condition is clearly not necessary, since if M has a canonical factorization, so does $-M$, for instance.) In other words, the LMI (7.2) holding for some $\epsilon > 0$ is a sufficient condition for a transfer function $M(s)$ with a state-space realization (A, B, C, D) to have a canonical factorization.
The question is: *Given (A, B, C, D), what, if any, are matrix inequality conditions that are both necessary and sufficient for the existence of a canonical factorization of $M(s) = C(sI - A)^{-1}B + D$?*

2 Motivations

Canonical factorization plays an important role in the analysis of systems with nonlinearities and/or uncertainties. A standard technique for the stability analysis of nonlinear and/or uncertain systems is the use of stability multipliers M, whose search is performed numerically [5, 6]. One of the conditions on M is that it should have a canonical factorization, and therefore simple numerical conditions that impose this constraint are important. Canonical factorization also figures in H_∞ control theory [7].

3 Available results

A necessary and sufficient condition for canonical factorization is given in [8] and [7, Ch. 7, Thm. 1]: M has a canonical factorization if and only if the invariant subspace of A corresponding to its eigenvalues with positive real part and that of $A - BD^{-1}C$ corresponding to its eigenvalues with negative real part are complementary. Though this condition is very explicit, it is somewhat hard to incorporate in numerical algorithms, and hence our objective to identify conditions that are easier to check numerically.

4 References

[1] D. C. Youla, "On the factorization of rational matrices," *IRE Trans. Information Theory*, vol. IT-7, pp. 172–189 (1961).

[2] B. Anderson and S. Vongpanitlerd, *Network analysis and synthesis: a modern systems theory approach.* Prentice-Hall, 1973.

[3] J. C. Willems, "Least squares stationary optimal control and the algebraic Riccati equation," *IEEE Trans. Aut. Control*, vol. AC-16, pp. 621–634 (1971).

[4] A. Rantzer, "A note on the Kalman-Yacubovich-Popov lemma," in *Proceedings of 3rd European Control Conference*, pp. 1792–1795, 1995.

[5] C. A. Desoer and M. Vidyasagar, *Feedback Systems: Input-Output Properties.* New York: Academic Press, 1975.

[6] V. Balakrishnan, "Linear matrix inequalities in robustness analysis with multipliers," *Syst. Control Letters*, vol. 25, no. 4, pp. 265–272 (1995).

[7] B. A. Francis, *A course in* $\mathbf{H}_\infty$ *Control Theory*, vol. 88 of *Lecture Notes in Control and Information Sciences.* Springer-Verlag, 1987.

[8] H. Bart, I. Gohberg, and M. A. Kaashoek, *Minimal Factorization of Matrix and Operator Functions.* Basel: Birkhauser, 1979.

8

Open problems in ℓ_1 optimal control

Bassam Bamieh* and Munther A. Dahleh**

*Department of Mechanical and Environmental Engineering
University of California at Santa Barbara
Santa Barbara
USA
bamieh@seidel.ece.ucsb.edu

**Laboratory for Information and Decision Systems
Massachusetts Institute of Technology
Cambridge, MA 02139
USA
dahleh@mit.edu

1 Description of the problem

The ℓ_∞ general disturbance rejection problem can be stated as follows: Find a stabilizing controller K that minimizes:

$$\sup_{\|w\|_\infty \leq 1} \|z\|_\infty \tag{8.1}$$

where z is the regulated variable and w is the disturbance input. If the plant controller are restricted to LTI systems, then the above quantity reduces to an ℓ_1 minimization of the closed loop map, and can equivalently be expressed using the Youla parametrization as:

$$\min_{Q \in \ell_1} \|H - UQV\|_1 \tag{8.2}$$

where H, U, V are all in ℓ_1. We will refer to Problem 1 when we are interested in general nonlinear controllers, and to Problem 2 when we focus only on linear controllers. The reader can consult [2, 3, 4, 1] for formulations and solutions of the above problems.

OP 1 It is well known that, under mild conditions, Problem 2 has a solution in the space ℓ_1. If H, U, V are all rational functions, does there exist a rational optimal solution? A rational optimal solution exists in the 1-block case; ($\hat{U}$ has full row normal rank and $\hat{V}$ has full column normal rank) and there are no counter examples for the multi-block cases.

OP 2 There is a weaker version of the problem stated above. In particular, can we find M and $\rho < 1$ such that all optimal solutions satisfy:

$$|\Phi(k)| \leq M\rho^k, \quad k = 0, 1, 2, \ldots$$

It turns out that such a bound can be derived for Problem 1 when we allow nonlinear state feedback controllers.

OP 3 It is well-known (see [2]) that 1-block problems are exactly equivalent to finite dimensional linear programs and hence can be solved exactly. This is not the case for multi-block problems. An efficient algorithm for solving such problems is known as the "Delay Augmentation" algorithm, which is based on augmenting the matrices $\hat{U}$ and $\hat{V}$ with delays to satisfy the necessary rank conditions for the 1-block problem. This adds additional degrees of freedom that are ultimately removed as the order of the delay term is increased. The convergence of this algorithm is studied in detail. Given a multi-block problem, a one block *partition* is obtained from taking a subset of the rows of U and the columns of V so that the resulting problem is 1-block. Such a partition is called *partially dominant* if all optimal ℓ_1 solutions are polynomials in the entries corresponding to this partition. It has been conjectured in [2] that the delay augmentation algorithm recovers all the partially dominant blocks of a given problem; i.e., the solution of the augmented problem will converge to a polynomial in the entries corresponding to a *partially dominant* block. This conjecture has been verified by many simulations, however, it remains an open problem.

OP 4 Problem 1 has been solved in the case of state feedback using dynamic programming and has been shown to result in nonlinear static controllers. The output feedback problem is still open. In addition, there is no known separation structure of these optimal controllers except in very limited cases ([5]).

OP 5 A related problem to the above ones is the model reduction problem. Given $G \in \ell_1$, what is the best r^{th} order system G_r that minimizes $\|G - G_r\|_1$? Let Γ_G be the standard Hankel operator on ℓ_∞. It follows trivially that $\|G\|_1 = \|\Gamma_G\|$. The above problem can be reduced to minimizing $\|\Gamma_G - \Gamma_{G_r}\|$. The advantage of the later problem is that Γ_{G_r} is a finite rank operator, and hence upper and lower bounds of this error can be derived. However, there is no known characterization of the optimal error.

2 References

[1] A. E. Barabanov, and A. A. Sokolov, "Geometrical Solution to ℓ_1-optimization Problem with Combined Conditions. In Proc. Asian Control Conference, vol. 3, pp. 331-334. Tokyo Japan, 1994.

[2] M. A. Dahleh, and I. J. Diaz-Bobillo, *Control of Uncertain Systems: A Linear Programming Approach*, Prentice-Hall , New Jersey, 1995.

[3] N. Elia and M.A. Dahleh, Minimization of the Worst-Case Peak to Peak Gain via Dynamic Programming: State Feedback Case, ACC 1997.

[4] J.S. Shamma, "Optimization of the ℓ_∞ induced norm under full state feedback", IEEE Transactions on Automatic Control, April 1996.

[5] J.S. Shamma, Set-valued observers and optimal disturbance rejection, submitted to IEEE Transactions on Automatic Control.

9

Efficient neural network learning

Peter L. Bartlett

Department of Systems Engineering
Research School of Information Sciences and Engineering
Australian National University
Canberra, ACT 0200
AUSTRALIA
Peter.Bartlett@anu.edu.au

1 Description of the problem

For a positive integer k, and parameters $a = (a_1, \ldots, a_k) \in \Re^k$ and $b = (b_1, \ldots, b_k) \in \Re^{dk}$, define the function

$$f_{k,a,b}(x) = \sum_{i=1}^{k} a_i \sigma(b_i \cdot x),$$

where $x \in \Re^d$, $b_i \cdot x$ denotes the inner product, and $\sigma : \Re \to \Re$ is the sigmoid function,

$$\sigma(\alpha) = \frac{1}{1+e^{-\alpha}}.$$

For A and B in $\Re^+ \cup \{\infty\}$, define the function class

$$F_{A,B} = \{f_{k,a,b} : k < \infty, \|a\|_1 < A, \|b_i\|_1 < B\},$$

where $\|a\|_1 = \sum_{j=1}^{d} |a_j|$. Then $F_{A,B}$ is the class of functions computed by a two-layer neural network with constraints on its parameters.
For $A, B, C, D \in \Re^+ \cup \{\infty\}$, $\epsilon \in \Re^+$, and positive integers d and n, define the following optimization problem.

$P(A, B, C, D, d, n, \epsilon)$: Given $x_1, \ldots, x_n$ in $(-C, C)^d$ and $y_1, \ldots, y_n$ in the interval $(-D, D)$, find k, a, b such that

$$\frac{1}{n}\sum_{i=1}^{n} (f_{k,a,b}(x_i) - y_i)^2 < \inf_{f \in F_{A,B}} \frac{1}{n}\sum_{i=1}^{n} (f(x_i) - y_i)^2 + \epsilon.$$

1. Is there an algorithm to solve $P(A, \infty, \infty, D, d, n, \epsilon)$ that runs in time polynomial in A, D, d, n, and $1/\epsilon$?

2. Is there an algorithm to solve $P(A, B, C, D, d, n, \epsilon)$ that runs in time polynomial in A, B, C, D, d, n, and $1/\epsilon$?

The same questions are of interest when σ is replaced by other nonlinear functions, such as the unit step function, or a bounded, non-constant, piecewise polynomial function.

2 History and motivation

Neural networks have been applied with some success to prediction and system modelling problems, typically using gradient descent techniques for parameter optimization. It is not surprising that these techniques suffer from the existence of non-global minima in the error surface; results in [10, 2] show that there can be many local minima of the error function, even if the number k of nonlinear functions is fixed at one (see also [9]). Results from computational learning theory show that a solution to the optimization problem $P(\cdots)$ gives guarantees on the performance of neural networks for nonlinear regression problems. In particular, if there is an efficient algorithm to solve $P(A, \infty, \infty, D, d, n, \epsilon)$, then the uniform convergence results in [7] imply that efficient agnostic pac learning of neural networks is possible. (In the *agnostic pac learning* framework, we assume that the data (X, Y) is i.i.d., and the aim is to find a function in the class $F_{A,\infty}$ with near minimal expected squared error. This learning model is studied in [1, 11].) Similarly, a positive answer to Question 2, together with results in [3], would imply the same result for the problem of efficient agnostic pac learning of $F_{A,B}$.
A second application is to the use of neural networks for nonlinear system output smoothing. For instance, we can use a neural network of the form $f_{k,a,b}(x)$ to model the output of an unknown SISO system in response to a certain input signal $(\ldots, u_{t-1}, u_t)$, where $x = (u_{t-d+1}, \ldots, u_t)$. Results in [4] show that the problem of near-optimal output noise smoothing reduces to this optimization problem of approximately minimizing squared error on training data.

3 Related results

Similar questions can be posed for networks without constraints on the size of the parameters, but with a limited number of parameters. In this case, Maass [8] has shown that the optimization problem can be solved in time polynomial in D, n, and $1/\epsilon$ (but not in the number of parameters) if the nonlinear function σ is replaced by a fixed piecewise polynomial function. If we require time polynomial in the number of parameters, the problem is known to be NP-hard for a variety of nonlinear functions σ (see [6] and related results in [5]). Recently, Vu has shown [12] that the optimization problem with a fixed number k of nonlinear functions is NP-hard for any fixed $\epsilon < ck^{-5/2}d^{-1/2}$, for some constant c. Without constraints on either the size or number of parameters, the optimization problem becomes trivial, but is not useful for learning.

If the function σ is replaced by a step function, the class $F_{A,\infty}$ is essentially unchanged, but Questions 1 and 2 become equivalent. In that case, Lee *et al* [7] have shown that the optimization problem can be efficiently solved if the *fan-in* of the input parameter vectors b_i is bounded by a constant. (The fan-in of $b_i = (b_{i,1}, \ldots, b_{i,d}) \in \Re^d$ is the number of non-zero components $b_{i,j}$.) Clearly, this implies a positive answer to Questions 1 and 2 if we remove the requirement that the computation time should grow polynomially in d.

4 References

[1] M. Anthony and P. Bartlett. *A Theory of Learning in Artificial Neural Networks.* (in preparation), 1998.

[2] P. Auer, M. Herbster, and M. Warmuth. Exponentially many local minima for single neurons. In *Advances in Neural Information Processing Systems, 8*, pages 316–322. MIT Press, 1996.

[3] P. L. Bartlett. The sample complexity of pattern classification with neural networks: the size of the weights is more important than the size of the network. *IEEE Transactions on Information Theory*, to appear, March 1998.

[4] P. L. Bartlett and S. Kulkarni. The complexity of model classes, and smoothing noisy data. *Systems and Control Letters*, to appear, 1998.

[5] A. L. Blum and R. L. Rivest. Training a 3-node neural network is np-complete. *Neural Networks*, 5:117–127, 1992.

[6] L. K. Jones. The computational intractability of training sigmoidal neural networks. *IEEE Transactions on Information Theory*, 43(1):167–173, 1997.

[7] W. S. Lee, P. L. Bartlett, and R. C. Williamson. Efficient agnostic learning of neural networks with bounded fan-in. *IEEE Transactions on Information Theory*, 42(6):2118–2132, 1996.

[8] W. Maass. Agnostic PAC-learning of functions on analog neural networks. *Neural Computation*, 7(5):1054–1078, 1995.

[9] E. D. Sontag. Critical points for least-squares problems involving certain analytic functions, with applications to sigmoidal nets. *Advances in Computational Mathematics*, 5:245–268, 1996.

[10] E. D. Sontag and H. J. Sussmann. Backpropagation can give rise to spurious local minima even for networks without hidden layers. *Complex Systems*, 3:91–106, 1989.

[11] M. Vidyasagar. *A Theory of Learning and Generalization*. Springer, 1997.

[12] V. H. Vu. On the infeasibility of training neural networks with small squared error. In *Advances in Neural Information Processing Systems, 10*. MIT Press, 1998. to appear.

10
Mechanical feedback control systems

Anthony M. Bloch*, Naomi Ehrich Leonard** and Jerrold E. Marsden***

*Department of Mathematics
University of Michigan
Ann Arbor, MI 48109
USA
abloch@math.lsa.umich.edu

**Department of Mechanical & Aerospace Engineering
Princeton University
Princeton, NJ 08544
USA
naomi@princeton.edu

***Control and Dynamical Systems 107-81
California Institute of Technology
Pasadena, CA 91125
USA
marsden@cds.caltech.edu

1 Description of the problem

We discuss here the problem of analyzing which classes of control systems may be put in a closed-loop form which is Lagrangian or Hamiltonian and ask when such controls are useful.

2 Motivation and history of the problem

In recent work (Bloch, Leonard and Marsden [1997, 1998], and Bloch, Marsden and Sánchez [1997]), the authors have analyzed a fairly large class of controlled mechanical (Lagrangian or Hamiltonian) systems which admit feedback controls which leave the system in a Lagrangian form. The feedback controlled systems are, in fact, in Lagrangian form with respect to a modified Lagrangian (which reflects the fact that the physical energy may be being pumped in or removed). Further, the controls preserve any natural symmetries of the system, leading to the existence of modified conservation laws via Noether's theorem. These conserved quantities enable one to achieve certain stabilization results.

The technique in our work has many features in common with the Kaluza-Klein theory and involves a modification of the system kinetic energy.[1] The way in which this is done is indicated below in equation (10.1). This work was inspired by earlier work: Krishnaprasad [1985], Bloch and Marsden [1990], Bloch, Krishnaprasad, Marsden and Sánchez [1992], and Wang and Krishnaprasad [1992], the last being of particular interest for the way *gyroscopic* effects were handled.

Hamiltonian stability is necessarily nonasymptotic, so in order to achieve asymptotic stability we first shape the energy and then introduce a suitable (usually active) dissipation. In our work so far we simply introduce dissipation at the linear level, but in other work we have considered the notion of nonlinear dissipation in certain settings (see Bloch, Krishnaprasad, Marsden and Ratiu [1996]). The reason for this is to formulate a dissipative mechanism which preserves quantities like angular momentum yet dissipates energy. An example of this is viscous fluid damping in satellites that extracts energy but not angular momentum. In the linear setting we think of dissipation as arising from a Rayleigh dissipation function and in the paper just mentioned this notion was generalized. An interesting early reference on dissipation in physical systems is the paper Willems [1979] (see also related references therein).

There has been other work on controlled Hamiltonian systems which also retain their Hamiltonian form. For example, the work of van der Schaft (see van der Schaft [1986]) considers modifying the system potential. Related work includes that of Koditschek [1989] and Leonard [1997].

A related but different problem is that of understanding the notion of controlled Hamiltonian or Lagrangian systems and deciding which control systems are of this type. (The latter question is related to the classical inverse problem in the calculus of variations.) In this setting one considers the open-loop system where the Hamiltonian or Lagrangian is a function of the controls u. A key paper which inspired much of the work of Hamiltonian control systems was that of Brockett [1976] – see also Brockett and Rahimi [1972], Brockett [1978], Hermann [1977], Willems [1979], van der Schaft

[1]Kaluza-Klein theory is a large subject, but here we are referring to a simple version of the procedure; an example concerns the dynamics of a charged particle moving in $\mathbf{R}^3$ in a given magnetic field. The Kaluza-Klein construction shows how solutions of this system become geodesics on a larger *unreduced* space, in this case $\mathbf{R}^3 \times S^1$. This larger space is a space in which an additional cyclic variable has been added; its conjugate momentum is the *charge* and the metric has been modified in a certain way. This simple example is described in Marsden and Ratiu [1994].

[1982], Crouch and van der Schaft [1987], Slotine [1988], Ortega et al. [1995], Lewis and Murray [1997] and the recent survey of Bloch and Crouch [1998]. The notion of constructing Hamiltonian systems with symmetry also occurs in circuit theory.

3 Available results and desired extensions

In the recent work of the authors, we assume we are given a mechanical Lagrangian of the form kinetic energy (given by a metric g) minus potential energy V defined on a velocity phase space TQ, with certain controls. We then define horizontal and vertical spaces at each point of TQ by modifying the mechanical connection (see Marsden [1982]) by a one-form τ which we relate to the controls. We then define a *controlled Lagrangian* by modifying the kinetic energy. This new Lagrangian is of the form

$$L_{\tau,\sigma,\rho}(v) = \frac{1}{2}\left[g_\sigma(\mathrm{Hor}_\tau v_q, \mathrm{Hor}_\tau v_q) + g_\rho(\mathrm{Ver}_\tau v_q, \mathrm{Ver}_\tau v_q)\right] - V(q). \quad (10.1)$$

The modified (controlled) Lagrangian is thus obtained by firstly choosing new horizontal and vertical spaces and then modifying the metric on each of these spaces. (Note that for τ chosen as the mechanical connection and the metrics g_σ and g_ρ set equal to the original metric g, the Lagrangian above reduces to the original Lagrangian.)

The parameters in the controlled Lagrangian are chosen to give a Lagrangian which reproduces the equations of the feedback controlled system. In addition, the controls are chosen to preserve natural symmetries in the system. In this fashion, one can find a class of energy and momentum based Lyapunov functions which can be used to prove closed-loop stability results. In the first instance one obtains nonlinear stability, but with the addition of dissipation, as mentioned above, one can get asymptotic stability. We have been able to apply this to such systems as the nonlinear pendulum on a cart, the spherical pendulum on a cart, and satellites and underwater vehicles. We also note that the techniques used are similar in spirit to those of Aström and Furuta [1996], which achieve efficient swing up and stability of a pendulum on a rotor arm.

Many interesting phenomena require the use of control ideas in the context of more complex dynamics such as periodic motions and heteroclinic orbits. Often the most interesting controls are those that utilize natural instabilities of components of a more general feature one wishes to stabilize. An example is a heteroclinic cycle connecting unstable saddle points. This type of control was explored in Bloch and Marsden [1989]. Similar ideas are being considered for underwater vehicle control (see Holmes, Jenkins and Leonard [1998]) and are being used in the mission planning for the NASA/JPL Genesis project to be launched in 2001.

The problem we pose is:

(a) Find other techniques which enable one to use feedback control for mechanical or, indeed, nonmechanical systems, which leave or put the system into Hamiltonian or Lagrangian form.

(b) Describe what analytic or performance advantages there are in having a feedback controlled system in Hamiltonian or Lagrangian form.

(c) Understand the additional role of dissipation for such systems and in particular the notion of nonlinear Rayleigh dissipation function.

(d) Define other qualitative properties of the physical systems that are preserved or modified in a useful fashion under feedback.

(e) Develop energy-based methods for stabilizing more general dynamic features than relative equilibria, such as periodic motions, heteroclinic connections and bifurcations.

Acknowledgments

We would like to thank R. Brockett for useful comments on this note. AMB would like to acknowledge partial research support by the National Science Foundation PYI grant DMS–9496221 and AFOSR grant F49620-96-1-0100, NEL acknowledges partial support from the National Science Foundation under grant BES-9502477 and by the Office of Naval Research under grant N00014-96-1-0052 and the research of JEM was partially supported by AFOSR Grant F49620-95-1-0419.

4 References

Aström, K.J. and K. Furuta [1996] Swinging up a pendulum by energy control. *IFAC*, **13**.

Bloch, A.M., and P. Crouch [1998] Optimal control, optimization and analytical mechanics, to appear in the *Brockettfest.*

Bloch, A.M., P.S. Krishnaprasad, J.E. Marsden, and T.S. Ratiu [1996] The Euler-Poincaré equations and double bracket dissipation. *Comm. Math. Phys.* **175**, 1–42.

Bloch, A.M., P.S. Krishnaprasad, J.E. Marsden and G. Sánchez de Alvarez [1992] Stabilization of rigid body dynamics by internal and external torques. *Automatica* **28**, 745–756.

Bloch, A.M. and J. Marsden [1989] Controlling homoclinic orbits, *Theoretical and Computational Fluid Dynamics* **1** (1989), 179–190.

Bloch, A.M. and J. Marsden [1990] Stabilization of the rigid body equations and the Energy–Casimir method, *Systems and Controls Letters* **14**, 341–346.

Bloch, A.M., J.E. Marsden and G. Sánchez [1997] Stabilization of relative equilibria of mechanical systems with symmetry, *Current and Future Directions in Applied Mathematics*, Edited by M. Alber, B. Hu, and J. Rosenthal, Birkhäuser, 43–64.

Bloch, A.M., N.E. Leonard and J.E. Marsden [1997] Stabilization of mechanical systems using controlled Lagrangians. *Proc. IEEE Conf. Dec. Contr.*, San Diego, CA, 2356–2361.

Bloch, A.M., N.E. Leonard and J.E. Marsden [1998] Controlled Lagrangians and the stabilization of mechanical systems. Preprint.

Brockett, R.W. [1976] Control theory and analytical mechanics, in *1976 Ames Research Center (NASA) Conference on Geometric Control Theory*, R. Hermann and C. Martin, eds., Lie Groups: History Frontiers and Applications, Vol. 7, Math. Sci. Press, Brookline, Mass., USA.

Brockett, R.W. [1978] Lie algebras and rational functions: some control theoretic connections, in *Lie Theories and Their Applications* (W. Rossman, ed.). Kingston, Ontario: Queen's University, Dept. of Mathematics, 268–280.

Brockett, R.W. and A. Rahimi [1972] Lie algebras and linear differential equations, in *Ordinary Differential Equations* (L. Weiss, ed.), Academic Press, 379–386.

Crouch, P. and A.J. van der Schaft [1987] *Variational and Hamiltonian Control Systems*, Lecture Notes in Control and Informations Sciences **10**, Springer Verlag.

Hermann, R. [1977] *Differential Geometry and the Calculus of Variations*, 2nd Edition, Interdisciplinary Mathematics Volume, **XVII**, Math. Sci. Press, Brookline, Mass., USA (First Edition 1968, Academic Press).

Holmes, P., J. Jenkins and N.E. Leonard [1998] Dynamics of Kirchhoff Equations I: Coincident Centers of Gravity and Buoyancy, *Physica D.* To appear.

Koditschek, D.E. [1989] The application of total energy as a Lyapunov function for mechanical control systems, in Dynamics and control of multibody systems (Brunswick, ME, 1988), 131–157, *Contemp. Math.*, **97**, Amer. Math. Soc., Providence, RI.

Krishnaprasad P.S. [1985] Lie-Poisson structures, dual-spin spacecraft and asymptotic stability, *Nonl. Anal. Th. Meth. and Appl.* **9**, 1011–1035.

Leonard, N.E. [1997] Stabilization of underwater vehicle dynamics with symmetry-breaking potentials, *Systems and Control Letters* **32**, 35–42.

Lewis, A.D. and R. Murray [1997] Configuration controllability of simple mechanical control systems, *SIAM Journal on Control and Optimization* **35**, 766-790.

Marsden, J.E. [1992] *Lectures on Mechanics*, London Mathematical Society Lecture note series. **174**, Cambridge University Press.

Marsden, J.E. and T.S. Ratiu [1994] *Introduction to Mechanics and Symmetry.* Texts in Applied Mathematics, **17**, Springer-Verlag.

Ortega, R. A. Loria, R. Kelly, and L. Praly [1995], On passivity-based output feedback global stabilization of Euler-Lagrange systems, *Int. J. Robust and Nonlinear Control*, special issue on Control of mechanical systems, **5** no 4, 313-325.

Slotine, J.-J. [1988] Putting physics in control: the example of robotics *IEEE Control Systems Magazine* **8**, 12-18.

Van der Schaft, A.J. [1982] Hamiltonian dynamics with external forces and observations, *Mathematical Systems Theory* **15**, 145–168.

Van der Schaft, A. J. [1986] Stabilization of Hamiltonian systems, *Nonlinear Analysis, Theory, Methods and Applications*, **10**, 1021–1035.

Wang, L.S. and P.S. Krishnaprasad [1992] Gyroscopic control and stabilization, *J. Nonlinear Science* **2**, 367–415.

Willems, J.C. [1979] System theoretic models for the analysis of physical systems, in *Ricerche di Automatica* **10**, Special Issue on Systems Theory and Physics (R.W. Brockett ed.) 71–106.

11

Three problems on the decidability and complexity of stability

Vincent D. Blondel* and John N. Tsitsiklis**

*Institute of Mathematics
University of Liège
Sart Tilman B37
4000 Liège
BELGIUM
vblondel@ulg.ac.be

**Laboratory for Information and Decision Systems
Massachusetts Institute of Technology
Cambridge, MA 02139
USA
jnt@mit.edu

1 Decidability and complexity

We describe three simply-stated problems that deal with the notion of stability.

All three problems are yes-no *decision* problems; upon input of the data associated with an instance of the problem, we wish to *decide* whether a certain property is satisfied by the instance. Many results are available in the literature for these three problems, but no satisfactory answers are yet available. We suggest looking at the decidability and at the computational complexity of these three problems.

We say that a problem is *decidable* if there is an algorithm which, upon input of the data associated with an instance of the problem, provides a yes-no answer after finitely many steps. The precise definition of what is meant by an *algorithm* is not critical; most algorithm models proposed so far are known to be equivalent from the point of view of their computing capabilities (see [Hopcroft and Ullman, 1969]).

We say that a problem can be decided in *polynomial time*, or that it can be decided *efficiently*, if there is a polynomial p and an algorithm which, upon input of an instance Σ of the problem, provides a yes-no answer after at most $p(size(\Sigma))$ computational steps. Again, the precise definition of $size(\Sigma)$, and of *computational step* are not critical. The property of being decidable in polynomial time is robust across all reasonable definitions. The class P is the class of problems that can be decided in polynomial time.
The class NP is a class of problems that includes all problems in P and includes a large number of problems of practical interest for which no polynomial time algorithms have yet been found. It is widely believed that $P \neq NP$. A problem is NP-hard if it is at least as hard as any problem in NP. A polynomial time algorithm for an NP-hard problem would immediately result in polynomial time algorithms for all problems in NP. Finally, a problem is NP-complete if it is NP-hard and belongs to NP. See [Papadimitriou, 1994] for more details. The reference [Blondel and Tsitsiklis, 1998] surveys complexity results available for systems and control problems.

2 Static output feedback

We are given an input-output linear system

$$\begin{aligned} \dot{x} &= Ax + Bu \\ y &= Cx \end{aligned}$$

and we consider a static feedback control law $u = Ky$. The resulting closed loop is

$$\dot{x} = (A + BKC)x.$$

The problem is to find conditions on the triplet of real matrices (A, B, C) under which there exists a feedback gain matrix K such that $A + BKC$ is stable, i.e., has all its eigenvalues in the left half plane.

STABILIZATION BY STATIC OUTPUT FEEDBACK.
Instance: Matrices A, B and C.
Problem: Does there exists a matrix K such that $A + BKC$ has all its eigenvalues in the left half plane?

Let n be the dimension of A. In the case of full state feedback ($C = I$) a necessary and sufficient condition for the the system to be stabilizable by static output feedback is that the rank of the matrix

$$(B, AB, A^2B, \ldots, A^{n-1}B)$$

is equal to n. This condition can be checked in a number of operations that is polynomial in the dimension of the matrices A and B; see, e.g., [Schrijver, 1986]. When C is invertible, a similar condition can be obtained easily. When C is not invertible, no general tractable necessary and sufficient conditions are known. After more than two decades of research it seems now unlikely that a closed-form solution exists to this problem. In [Anderson, Bose and Jury, 1975] it is shown that an *algorithmic* solution is possible. The algorithm proposed in this reference is based on the Tarski-Seidenberg elimination algorithm and uses, in the worst case, a number of operations that grows faster than any polynomial in the number of input and output variables.

Open Problem 1: Can STABILIZATION BY STATIC OUTPUT FEEDBACK be solved in time polynomial in the size of the matrices A, B and C? Is the problem NP-hard?

In [Blondel and Tsitsiklis, 1997] it is shown that the following related *constrained* problem is indeed NP-hard.

STABILIZATION BY CONSTRAINED STATIC OUTPUT FEEDBACK.
Instance: Matrices A, B and C, rational numbers $\underline{k}_{ij}, \overline{k}_{ij}$.
Problem: Does there exists a matrix $K = (k_{ij})$ satisfying $\underline{k}_{ij} \leq k_{ij} \leq \overline{k}_{ij}$ and such that $A + BKC$ has all its eigenvalues in the left half plane?

There does not seem to be any easy extension of the proof of this result for the unconstrained case.

3 Stability of all infinite products

Let $\Omega = \{A_1, \ldots, A_m\}$ be a set of $n \times n$ real matrices. Given a system of the form

$$x_{t+1} = A_t x_t \tag{11.1}$$

suppose that it is known that $A_t \in \Omega$, for each t, but that the exact value of A_t is not a priori known, because of exogenous conditions or changes in the operating point of the system. This system can also be thought of as a time-varying system. We say that such a system is *stable* if

$$\lim_{t \to \infty} x_t = 0$$

for all initial states x_0 and all sequences of matrix products. This condition is equivalent to the requirement

$$\lim_{t \to \infty} A_{i_t} \cdots A_{i_1} A_{i_0} = 0$$

for all sequences of indices i_j.

STABILITY OF ALL INFINITE PRODUCTS.
Instance: A finite set of $n \times n$ matrices $\Omega = \{A_1, \ldots, A_m\}$.
Problem: Do the products

$$A_{i_t} \cdots A_{i_1} A_{i_0}$$

converge to zero for all sequences of indices i_j?

This problem is obviously decidable when $n = 1$ and when $m = 1$. No general decision algorithms are known for any other values of n and m.

Open Problem 2: For what values of n and m is STABILITY OF ALL INFINITE PRODUCTS decidable?

The problem is known to be related to the *finiteness conjecture* on the generalized spectral radius of matrices. Let $\rho(A)$ denote the *spectral radius* of a real matrix A,

$$\rho(A) := \max\{|\lambda| : \lambda \text{ is an eigenvalue of } A\}.$$

Let $\Omega = \{A_1, \ldots, A_m\}$ be a finite set of matrices. The *generalized spectral radius* $\rho'(\Omega)$ is defined in [Daubechies and Lagarias, 1992] by

$$\rho'(\Omega) = \limsup_{k \to \infty} \rho'_k(\Omega), \tag{11.2}$$

where

$$\rho'_k(\Omega) = \max\{(\rho(A_1 A_2 \cdots A_k))^{1/k} : \text{each } A_i \in \Omega\}$$

for each $k \geq 1$. It is conjectured in [Lagarias and Wang, 1995] that the equality $\rho'(\Omega) = \rho'_k(\Omega)$ always occur for some finite k. If this conjecture is true, then STABILITY OF ALL INFINITE PRODUCTS is decidable. Conversely, if the problem is undecidable for some n and m, then the finiteness conjecture must be false.

The related problem for which the matrices in (11.1) occur with a certain probability is studied in [Tsitsiklis and Blondel, 1997]. The corresponding stability problem is then undecidable. See also [Gurvits, 1995].

4 Stability of systems of the neural type

Let $\sigma : \mathbf{R} \mapsto \mathbf{R}$ be a fixed scalar function and consider the system

$$x_{t+1} = \sigma(A x_t) \tag{11.3}$$

where σ is defined componentwise and A is a real matrix. The system is *stable* if

$$\lim_{t\to\infty} x_t = 0$$

for all initial states x_0.

STABILITY OF σ-SYSTEMS.
Instance: A $n \times n$ matrices A.
Problem: Is the system $x_{t+1} = \sigma(Ax_t)$ stable?

The dynamics of such systems depends on the function σ. When σ is linear, the system is linear and stability is easy to check. When σ has finite range, stability can be decided by simple enumeration since there are only finitely many possible states.

Open problem 3: What are the functions σ for which stability of $x_{t+1} = \sigma(Ax_t)$ is undecidable?

The systems $x_{t+1} = \sigma(Ax_t)$ arise in a wide variety of situations. In particular, recurrent artificial neural networks are modeled by such equations where the function σ is the activation function used in the network. In [Siegelmann and Sontag, 1991] it is shown that, when σ is the saturated linear function, systems of this type are capable of simulating arbitrary Turing machines. Thus, as computational devices, linear saturated systems are as powerful as Turing machines. From this result it is easy to prove that the problem of deciding whether a given initial state of a saturated linear system eventually reaches a certain state (that encodes a halting configuration), is undecidable (see [Sontag, 1995]). Similar simulations are given in [Koiran, 1996] for a large class of other functions σ. These results do however not have direct implications for the decidability of stability of such systems. Undecidability of stability for saturated linear systems was conjectured in [Sontag, 1995].

Acknowledgments

This research was sponsored by the European Commission (TMR (Alapedes), network contract ERBFMRXCT960074) and by the NATO under grant CRG-961115.

5 References

[Anderson, Bose and Jury, 1975] Anderson B. D. O., N. K. Bose and E. I. Jury (1975). *Output feedback stabilization and related problem – solutions via decision methods*, IEEE Trans. Autom. Control, **20**, 53–66.

[Blondel and Tsitsiklis, 1997] Blondel, V. D. and J. N. Tsitsiklis (1997). NP-hardness of some linear control design problems, *SIAM J. Control and Optimization*, **35**, 2118-2127.

[Blondel and Tsitsiklis, 1997] Blondel V. D. and J. N. Tsitsiklis (1997). Complexity of stability and controllability of elementary hybrid systems. Report LIDS-P 2388, LIDS, Massachusetts Institute of Technology, Cambridge, MA, 1997. To appear in *Automatica*.

[Blondel and Tsitsiklis, 1998] Blondel, V. D. and J. N. Tsitsiklis (1998). Survey of complexity results for systems and control problems, (in preparation).

[Daubechies and Lagarias, 1992] I. Daubechies and J. C. Lagarias (1992), Sets of matrices all infinite products of which converge, Linear Algebra Appl., **162** 227-263.

[Gurvits, 1995] L. Gurvits (1995), Stability of discrete linear inclusion, Linear Algebra Appl., **231** 47-85.

[Hopcroft and Ullman, 1969] Hopcroft, J. E. and J. D. Ullman (1969). *Formal languages and their relation to automata*, Addison-Wesley.

[Koiran, 1996] Koiran, P. (1996). *A family of universal recurrent networks,* Theor. Comp. Science, **168**, 473-480.

[Lagarias and Wang, 1995] J. C. Lagarias and Y. Wang (1995), *The finiteness conjecture for the generalized spectral radius of a set of matrices,* Linear Algebra Appl., **214** 17-42.

[Papadimitriou, 1994] Papadimitriou, C. H. (1994). *Computational complexity*, Addison-Wesley, Reading.

[Siegelmann and Sontag, 1991] Siegelmann, H. and E. Sontag (1991). *Turing computability with neural nets*, Appl. Math. Letters, **4**, 77-80.

[Siegelmann and Sontag, 1995] Siegelmann, H. and E. Sontag (1995). *On the computational power of neural nets*, J. Comp. Syst. Sci., 132–150.

[Schrijver, 1986] Schrijver, A. (1986). *Theory of linear and integer programming*, Wiley.

[Sontag, 1995] Sontag, E. (1995). *From linear to nonlinear: some complexity comparisons*, Proc. IEEE Conference Decision and Control, New Orleans, 2916–2920.

[Tsitsiklis, 1987] Tsitsiklis, J. N (1987). The stability of the products of a finite set of matrices, in *Open problems in communication and computation*, Springer-Verlag.

[Tsitsiklis and Blondel, 1997] Tsitsiklis, J. N. and V. D. Blondel (1997). *Spectral quantities associated with pairs of matrices are hard – when not impossible – to compute and to approximate*, Mathematics of Control, Signals, and Systems, **10**, 31-40.

12

Simultaneous stabilization of linear systems and interpolation with rational functions

Vincent D. Blondel

Institute of Mathematics
University of Liège
Sart Tilman B37
4000 Liège
BELGIUM
vblondel@ulg.ac.be

1 Description of the problem

Although champagne has been offered for its solution, the following simple-looking problem is still unsolved (see below for more details):

For what values of δ is the system

$$p_\delta(s) = \frac{s^2 - 1}{s^2 - 2\delta s + 1}$$

stabilizable by a stable controller whose inverse is also stable?

In this contribution we describe the *simultaneous stabilization problem* and outline its connection with interpolation by rational functions and with other stabilization problems.

A controller c *simultaneously stabilizes* the systems $p_1, p_2, \ldots, p_k$ if it stabilizes all the systems. Tractable conditions for two systems to be simultaneously stabilizable are given in [13]. Let $\mathbf{R}(s)$ denote the set of real rational functions. Two systems $p_1 \in \mathbf{R}(s)$ and $p_2 \in \mathbf{R}(s)$ that have no common unstable real poles are simultaneously stabilizable if and only if the difference system $p := p_1 - p_2 \in \mathbf{R}(s)$ is stabilizable by a *stable* controller. Furthermore, as shown in [15], a system is stabilizable by a stable controller if and only if it has an even number of real unstable poles between every pair of real unstable zeros. Contrary to the case of two systems there are yet no tractable necessary and sufficient conditions available for simultaneous stabilization of three (or more) systems.

Open problem 1 (control version): Under what conditions are three linear systems simultaneously stabilizable?

This problem constitutes in some sense the most elementary robust control design problem. Practical motivations for considering this problem are numerous and can be found, e.g., in [2] and [10]. Although simply-stated the problem is difficult. In particular, as explained below, a complete solution to the problem would provide, as a side result, an answer to an unsolved question on interpolation by rational functions.

2 Stabilization and intersection

Let $\mathbf{C}_\infty = \mathbf{C} \cup \{\infty\}$ denote the extended complex plane and let $D = \{z \in \mathbf{C} : |z| \leq 1\}$ denote the closed unit disk. We say that a rational function $q \in \mathbf{R}(s)$ is *stable* if it has no poles in D. The function is *bistable* if q and q^{-1} are both stable. Adaptation of the problems and results that follow to instability regions different from D is an easy exercise.

The *degree* of a rational function is the largest of its polynomial numerator and denominator degrees. The *relative degree* is equal to the difference between the denominator degree and the numerator degree, it is also equal to the number of poles of the function at infinity. Rational function with positive relative degree are said to be *proper*. In practice, systems are always proper and controllers are required to be proper. It is known that systems that are simultaneously stabilizable by a (non-necessarily proper) controller are also simultaneously stabilizable by a proper controller; see Theorem 3.5 in [2]. Here we are only concerned by existence conditions for simultaneous stabilization and we therefore evacuate the properness issue.

Rational functions take points from $\mathbf{C}_\infty$ to values in $\mathbf{C}_\infty$. Let n be the degree of the rational function $p \in \mathbf{R}(s)$, the image of $\mathbf{C}_\infty$ under p covers the extended complex plane $\mathbf{C}_\infty$ exactly n times. In particular, a rational function of order n has exactly n poles and n zeros. The *intersections* between $p \in \mathbf{R}(s)$ and $q \in \mathbf{R}(s)$ are the points $s_0 \in \mathbf{C}_\infty$ for which

$$p(s_0) = q(s_0).$$

In particular, two rational function that have a common pole intersect there. Rational functions of respective degrees n_p and n_q have exactly $n_p + n_q$ points of intersection. (When counting intersections, multiplicities of the intersections need to be taken into account.) In the sequel we always assume that the intersections are simple.

There exists a simple relation between intersection and stabilization: The system p is stabilized by the controller c if and only if the rational functions p and $-c^{-1}$ have no intersections in D: see [3]. The simultaneous stabilization problem can thus be rephrazed as follows.

Open problem 1 (interpolation version): Under what conditions on $p_1, p_2, p_3 \in \mathbf{R}(s)$ does there exists a rational function $q \in \mathbf{R}(s)$ such that

$$q(s) \neq p_i(s)$$

for $i = 1, 2, 3$ and $s \in D$?

The problem of simultaneous stabilization thus has an easily understandable geometric interpretation. We are given a finite set of rational functions and we search for a rational function that does not intersect with these functions in the instability region. In the next section we describe a special case of this problem that is of independent interest.

3 Bistable stabilization

Simultaneous stabilization of two systems is equivalent to stabilization of a single system by a stable controller. This idea can, modulo an avoidance condition, be extended to the simultaneous stabilization of three systems.

Suppose that two of the systems $p_1, p_2, p_3 \in \mathbf{R}(s)$ do not intersect in D. Then one can construct a system p from p_1, p_2, p_3 (see [2] for an explicit construction) such that p_1, p_2, p_3 are simultaneously stabilizable if and only if p is stabilizable by a *bistable* controller.

The problem of finding conditions under which a system can be stabilized by a bistable controller can thus be seen as an intermediate step towards the solution of the simultaneous stabilization problem for three systems.

Open problem 2 (control version): Under what condition can a system be stabilized by a controller that is both stable and has a stable inverse?

It is known that a system p is stabilizable by a bistable controller if and only if there exists a stable rational function q satisfying

$$\begin{aligned} q(s) = 0 \quad &\Leftrightarrow \quad p(s) = 0 \\ q(s) = 1 \quad &\Leftrightarrow \quad p(s) = \infty \end{aligned}$$

for all s in the instability region. And thus, Open problem 2 can be rephrazed as follows.

Open problem 2 (interpolation version): Under what conditions on $Z = \{z_i \in D : i = 1, \ldots, n_z\}$ and $O = \{o_i \in D : i = 1, \ldots, n_o\}$ does there exists a stable rational function q for which $\{z \in D : q(z) = 0\} = Z$ and $\{z \in D : q(z) = 1\} = O$?

The cases $(n_z = 0, n_o \geq 0)$, $(n_z \geq 0, n_o = 0)$, and $(n_z = n_o = 1)$ always have a solution. As illustrated below with Problems 2 and 3, the case $(n_z = 1, n_o = 2)$ is unsolved.

4 Champagne problems

The next simply-stated particular stabilization problems have been proposed in the literature over the last five years with, in some cases, a reward for a solution. The problem appearing in [4] was proposed as an illustration of the difficulty of the simultaneous stabilization problem. The same problem is mentioned in [5] where a bottle of good French champagne is offered for its solution. This problem was solved in June 1998 by Dr. Vijay V. Patel. None of the other rewards have yet been claimed.

Problem 1: Can the continous-time second order system

$$p(s) = \frac{s^2 - 1}{s^2 - 1.8s + 1}$$

be stabilized by a stable controller whose inverse is also stable?
This problem appears on p.149 of [2]. Notice that there is a misprint in [2]. As it is clear from the context, the system on p.149 should read $(s^2-1)/(s^2 - 1.8s + 1)$, not $(s-1)^2/(s^2 - 1.8s + 1)$. One kilogram of world-famous Belgian chocolates are offered in [2] for a bistable controller, or for a proof that no such controller exists for this system.

Problem 2: Another kilogram of chocolate is offered in [2] for the more difficult problem of finding the *range* of δ for which the continuous-time system

$$p_\delta(s) = \frac{s^2 - 1}{s^2 - 2\delta s + 1}$$

is stabilizable by a stable controller whose inverse is also stable. A solution to this problem would, of course, lead to a solution to Problem 2. Let us note that, using the usual conformal mapping from the unit disk to the right half-plane, it is easy to see that the continuous-time system

$$p_\delta(s) = \frac{s^2 - 1}{s^2 - 2\delta s + 1}$$

is stabilizable by a bistable controller if and only if the discrete-time system

$$p_\delta(z) = \frac{z}{z^2 + \frac{1-\delta}{1+\delta}}$$

is stabilizable by a bistable controller. In [4] it is shown that there exists a positive constant β^* such that the system

$$\frac{z}{z^2 + \beta}$$

is stabilizable by a bistable controller when $\beta > \beta^*$, and is not stabilizable by such a controller when $\beta < \beta^*$. It is known that

$$10^{-5} < \beta^* < 1/e^2 = 0.1353\ldots.$$

The bounds $3.6\,10^{-10} < \beta^* < 1/e^2 = 0.1353\ldots$ were first given by R. Rupp. The lower bound $3.6\,10^{-10}$ is improved to $10^{-5} < \beta^*$ in [6] where it is also shown that a stable rational function the assumes the value 0 and 1 at least once and an unequal number of times in a disk of radius 10^{-5} centered at the origin must assume the value 0 or 1 elsewhere in the unit disk. Problem 2 is equivalent to that of finding the exact value of β^*. Problem 1 asks whether or not $\beta^* > (0.1)/1.9 = 0.0526$.

Problem 3: As an indication of the potential complexity of simultaneous stabilization questions, let us mention that the innocent looking discrete-time systems

$$p_1(z) = \frac{1}{z} \quad p_2(z) = \frac{1}{z} + \beta \quad p_3(z) = \frac{1}{z} - \beta$$

are simultaneously stabilizable if and only if

$$|\beta| < \frac{\Gamma^4(1/4)}{4\pi^2} = 4.377\ldots$$

where Γ is the gamma function; see [1] for a proof. Completely general conditions for the systems

$$p_1(z) = \frac{1}{z} + \beta_1 \quad p_2(z) = \frac{1}{z} + \beta_2 \quad p_3(z) = \frac{1}{z} + \beta_3$$

to be simultaneously stabilizable are given in [1]. These conditions are specific to these particular first order systems. We offer 2 bottles of whisky for general stabilizability conditions for three arbitrary *first order* systems. In addition to this, we double the rewards offered for a solution to Problem 3.

5 References

[1] D. Bertilsson, V. Blondel, Transcendence in simultaneous stabilization, J. of Math. Systems, Estimation, and Control, 6, 1-22, 1996.

[2] V. Blondel, Simultaneous stabilization of linear systems, Springer-Verlag, 1994.

[3] V. Blondel, G. Campion, M. Gevers, Avoidance and intersection in the complex plane, Proc. of the 30th Conference on Decision and Control, Brighton, UK, 47-48, 1991.

[4] V. Blondel, M. Gevers, R. Mortini, R. Rupp, Simultaneous stabilization of three or more systems: conditions on the real axis do not suffice, SIAM J. of Control and Optimization, 32, 2, 572-590, 1994.

[5] V. Blondel, M. Gevers, Simultaneous stabilization of three linear systems is rationally undecidable, Math. of Control, Signals, and Systems, 6, 135-145, 1994.

[6] V. Blondel, R. Rupp, H. Shapiro, On zero and one points of analytic functions, Complex Variables: Theory and Applications, 28, 189-192, 1995.

[7] B. Ghosh and C. Byrnes, Simultaneous stabilization and pole-placement by nonswitching dynamic compensation, *Trans. Automat. Contr.*, 28, pp. 735-741, 1983.

[8] B. Ghosh, Some new results on the simultaneous stabilizability of a family of single input single output systems, *Systems Control Lett.*, 6, pp. 39-45, 1985.

[9] B. Ghosh, Transcendental and interpolation methods in simultaneous stabilization and simultaneous partial pole placement problems, *SIAM J. Control and Optimiz.*, 24, pp. 1091-1109, 1986.

[10] B. Ghosh, An approach to simultaneous system design. Part 1, *SIAM J. Control and Optimiz.*, 24, pp. 480-496, 1986.

[11] B. Ghosh, An approach to simultaneous system design. Part 2, *SIAM J. Control and Optimiz.*, 26, pp. 919-963, 1988.

[12] R. Rupp, A covering theorem for a composite class of analytic functions, Complex Variables: Theory and Applications, 25, 35-41, 1994.

[13] R. Saeks and J. Murray, Fractional representation, algebraic geometry and the simultaneous stabilization problem, *IEEE Trans. Automat. Contr.*, 27, pp. 895-903, 1982.

[14] M. Vidyasagar, Control System Synthesis: A factorization approach, *MIT Press*, 1985.

[15] D. Youla, J. Bongiorno and C. Lu. Single-loop feedback stabilization of linear multivariable plants, *Automatica*, 10, pp. 159-173, 1974.

13
Forbidden state control synthesis for timed Petri net models

R.K. Boel and G. Stremersch[1]

Vakgroep Elektrische Energietechniek
University of Ghent
Technologiepark-Zwijnaarde
9052 Ghent
BELGIUM
Rene.Boel@rug.ac.be
Geert.Stremersch@rug.ac.be

1 Introduction

Complex, computer controlled plants can be analyzed efficiently by reducing the very large state space to a finite state space, using abstraction. At the same time the model can be decomposed in smaller components. Automata can be used as models for — components of — such discrete event systems. Proper behaviour of the system means that the global state of the system never reaches forbidden subsets, or equivalently, that certain forbidden sequences of transitions between states never occur. Ramadge and Wonham [6] developed a framework for control of discrete event systems. The state evolution can be constrained by blocking some controllable transitions. For a class of untimed Petri nets Holloway and Krogh [4] developed an efficient algorithm for synthesizing maximally permissive control laws, guaranteeing that the state never reaches some forbidden set. It turns out that this maximally permissive control law only depends on the marking of places in a subnet of the Petri net, and that control action is only required at transitions at the boundary of this same subnet. This subnet is called the influencing net [1, 4, 7].

---

[1]The results presented in this paper have been obtained within the framework of the Belgian Program on Interuniversity Attraction Poles, initiated by the Belgian State, Prime Minister's Office, Science Policy Programming. The scientific responsibility rests with its authors. R.K. Boel is supported by the Flemish Foundation for Scientific Research as Senior Research Associate.

Many applications of control of discrete event systems involve real time. Here we discuss the problem of extending the control synthesis algorithms for forbidden state control, to Petri nets which operate in real time. Control is achieved by the ability to delay controllable transitions within their allowed execution interval. Section 2 defines a controlled timed Petri net in detail and specifies its state. Section 3 gives a description of the forbidden state set and section 4 discusses which problems have to be overcome in order to extend the control synthesis algorithms based on decomposition of the net, from the untimed to the timed Petri net case.

2 Controlled timed Petri nets

A *Petri net* is a triple (P, T, F) with P a finite set of places, T a finite set of transitions and $F \subset (P \times T) \cup (T \times P)$ a set of relations. The sets P and T are disjoint. The state of a Petri net is its *marking* which is a function $m : P \rightarrow \mathbb{N}$, assigning to each place a nonnegative integer number of tokens. The set ${}^{\bullet}t := \{p \in P \mid (p,t) \in F\}$ ($t^{\bullet} := \{p \in P \mid (t,p) \in F\}$) contains the input (output) places of $t \in T$. The sets of transitions ${}^{\bullet}p$ and $p^{\bullet}$ for $p \in P$ are defined analogously. A transition t is called *state-enabled* under marking m if all its input places are marked, i.e. $\forall p \in {}^{\bullet}t : m(p) > 0$. When t is state-enabled then t can fire: One token is removed from each of the places in ${}^{\bullet}t$ and one token is added to each of the places in $t^{\bullet}$.
Define $\forall z_1, z_2 \in \mathbb{Z} \cup \{-\infty, \infty\}$ the set $[z_1, z_2] := \{z \in \mathbb{Z} \cup \{-\infty, \infty\} \mid z_1 \leq z \leq z_2\}$. A *timed Petri net* is a five-tuple (P, T, F, L, U) in which (P, T, F) is a Petri net. The functions L and $U : T \rightarrow \mathbb{N} \cup \{\infty\}$ are such that $\forall t \in T : L(t) \leq U(t)$. With each transition $t \in T$ we associate an integer time interval $[L(t), U(t)]$, the set of all possible time delays. If t becomes state-enabled — which is defined as in the untimed case — at time $\theta_e \in \mathbb{Z}$ then the transition can and must fire at some time $\theta_f \in [\theta_e + L(t), \theta_e + U(t)] \subset \mathbb{Z} \cup \{\infty\}$, provided t did not become disabled because of the firing of another transition. Notice that t is forced to fire if it is still state-enabled at $\theta_e + U(t)$. We only consider integer firing times. This is not a restriction: If they are rational, all execution times can be made integer by using as unit of time step the inverse of the smallest common multiple of all the denominators. The state of a timed Petri net at time $\theta \in \mathbb{Z}$ is a map $M_\theta : P \rightarrow \mathcal{L}_{[-\infty,\theta]}$ where $\mathcal{L}_{[-\infty,\theta]}$ denotes the set of all lists with elements in $[-\infty, \theta]$. If $p \in P$ contains $n \in \mathbb{N}$ tokens at θ, the list $M_\theta(p) := \{\tilde{\theta}_1, \ldots, \tilde{\theta}_n\}$ enumerates the arrival times of these n tokens in p. The set of all possible states is denoted by $\mathcal{M}$. The number of tokens in place p under marking M_θ is denoted by $m_\theta(p) := \sharp M_\theta(p)$. A transition t becomes state-enabled at time $\theta_e = \max_{p \in {}^{\bullet}t} \min M_{\theta_e}(p)$. By convention the minimum of the empty

list is equal to infinity. Transition t fires at some time

$$\theta_f \in [\max_{p\in{}^\bullet t} \min M_{\theta_f}(p) + L(t), \max_{p\in{}^\bullet t} \min M_{\theta_f}(p) + U(t)]. \qquad (13.1)$$

Execution of t at θ_f changes the distribution of tokens as in the untimed case. The state changes as follows: ¿From each place $p_1 \in {}^\bullet t$ the token which arrived there first, at $\min M_{\theta_f}(p_1)$, is removed from the list $M_{\theta_f}(p_1)$ and to each place $p_2 \in t^\bullet$ a token with value θ_f is added to the list $M_{\theta_f}(p_2)$. When interpreting such a model as a hybrid sytem, the firing of a transition at time θ_f in the timed Petri net corresponds to the occurrence of the corresponding event at time θ_f in the hybrid sytem. In case $\forall t \in T : U(t) = \infty$ then no transition is ever forced to fire, and the language of the timed Petri net is identical to the language generated by the untimed Petri net (P, T, F).

A *controlled timed Petri net* is a six-tuple (P, T, F, L, U, T^c) in which (P, T, F, L, U) is a timed Petri net. The set $T^c \subset T$ contains the controllable transitions. The firing of a controllable transition can be delayed. A control function C is a function $C : T^c \to \mathbb{N} \cup \{\infty\} : t \mapsto C(t)$ such that $L(t) \leq C(t) \leq U(t)$, changing the associated interval of possible firing times for transitions $t \in T^c$ from $[L(t), U(t)]$ to $[C(t), U(t)]$. This means that — in contradiction to the untimed case — the firing of a state-enabled controllable transition can not always be delayed forever. The set of all control functions is denoted by $\mathcal{C}$.

3 Problem formulation

In forbidden state feedback control for controlled timed Petri nets, the state of the net is not allowed to enter the set of forbidden markings $\mathcal{F} \subset \mathcal{M}$. Most realistic requirements, possibly after the net has been extended with extra places, can be modelled by $\mathcal{F} = \bigcup_{k=1}^{K} B_k$, where the constraint sets $B_k := \{M_\theta \in \mathcal{M} \mid a_k \leq \sum_{p\in P_k} c_{k,p}.m_\theta(p) \leq b_k\}$ only depend on the cardinality of the state. Places $p \in P_k \subset P$ are called forbidden places. The problem is to find a feedback control, if it exists, $C_f : \mathcal{M} \to \mathcal{C}$ which satisfies the control specifications $\mathcal{F}$. The set of all feedback controls is denoted by $\mathcal{C}_f$. Note that in case $\exists p \in P : \forall t \in {}^\bullet p : U(t) = \infty$, as for untimed models, we can never guarantee a minimum number of tokens in p. In that case necessarily $b_k = \infty$ and $c_{k,p} \geq 0$ for every constraint set B_k involving such a place p.

The control synthesis problem above defines a set of allowable solutions $\mathcal{C}_a \subset \mathcal{C}_f$. It is important that this set of allowable behaviours is made as large as possible. The set $\mathcal{C}_a$ is a partially ordered set under the natural ordering of control laws. Hence we can describe $\mathcal{C}_a$ via its maximal elements. Using such a maximally permissive control law leaves freedom to add further constraints, to minimize cost criteria, etc. Notice that, as

the upper bounds $U(t)$ are never changed, the system is not controlled by forcing the execution of a transition. However, the model is such that whenever some higher layer controller forces a transition t in the allowed interval $[C_f(M_\theta)(t), U(t)]$ the system will still behave according to the specification. The control synthesis algorithm may be interpreted as a tool for defining the set of allowable controllers, similar to the Youla parametrization for stable controllers of linear systems.
Assume that, if the initial marking is bounded, the structure of the Petri net implies that $\exists B \in \mathbb{N} : \forall p \in P, \forall \theta \in \mathbb{Z} : m_\theta(p) \leq B$. Note that we only need to define the states at the time just prior to the execution of a transition. Since the model is stationary, we can express all time values in the state representation as measured from the present time, i.e. at time θ we replace each element $\tilde{\theta}_j$ of $M_\theta(p)$ by $\tilde{\theta}_j - \theta$. No new information is conveyed by knowing that a transition t with $U(t) = \infty$ has been state-enabled for more than a time delay $L(t)$. Hence for each place there is an upper bound on the time delays we have to keep track of in the state M_θ, and the state space can be made bounded. This makes the problem decidable. But a solution by enumeration will inevitably be of extremely high complexity. In fact it is a special case of the CTL supervisor synthesis problem, known to be NP-complete.

4 Decomposition techniques for control synthesis

Efficient algorithms for solving the control synthesis problem will have to be based on a decomposition of the problem in simpler subproblems. To simplify the statement of the problem we assume that the constraint is of the simplest possible form which presents all the major difficulties of the problem: $\mathcal{F} = \{M_\theta \in \mathcal{M} \mid a \leq m_\theta(p) \leq b\}, a, b \in \mathbb{N}$. In the case of a controlled untimed Petri net — firing of state-enabled controllable transitions can be delayed arbitrarily long — the control synthesis algorithm requires only the observation of the marking of the influencing net $\mathcal{N}_p$ of the place p, i.e. the sub-Petri net formed by the set of places $p' \in P$ such that there exists an uncontrollable path from p' to p. It has been shown in [1, 4, 7] that there are efficient algorithms to decide which controlled transitions can be allowed to execute. These algorithms are worst case exponential in the number of choice places in $\mathcal{N}_p$, but this number is usually quite small compared to the size of the net. Special structure of $\mathcal{N}_p$ allows further reduction of the complexity: For example if $\mathcal{N}_p$ can be decomposed in a small number of state machines — all transitions have maximum one input and one output place — and marked graphs — all places have maximum one input and one output transition.
Maximally permissive allowable control laws for a controlled timed Petri net, with $U(t) < \infty$ for some t, will depend on the marking M_θ of a larger

subnet $\mathcal{N}_p^* \supset \mathcal{N}_p$. One has to take into account that control decisions cannot delay tokens by more than $U(t)$. $\mathcal{N}_p^*$ will include all places in those paths ending in p such that all transitions in the path are either uncontrollable or are controllable with a finite upper bound. Moreover knowledge is required about forced removal of tokens from $\mathcal{N}_p^*$. Hence $\mathcal{N}_p^*$ also contains all paths from where tokens can be forced to reach in a finite time any input place of an output transition t, with $U(t) < \infty$, of a place in $\mathcal{N}_p^*$. This takes into account the situation where a token can be forced out of p provided $\exists t \in p^\bullet : U(t) < \infty$. This makes the influencing net much larger. Control decisions now must be be taken both at the boundary of $\mathcal{N}_p^*$, allowing new tokens to enter $\mathcal{N}_p^*$ at some time instants, as well as inside the influencing net, delaying tokens already in $\mathcal{N}_p^*$.

In case the influencing net is a state machine all tokens, as tokens do not have an identity, move independently of each other. For each token in $\mathcal{N}_p^*$ at time θ, and for each token which will be allowed to enter $\mathcal{N}_p^*$ at some time after θ to be determined by the control decisions, one can calculate the finite union of intervals (one of them possibly unbounded), subset of $\mathbb{N} \cup \{\infty\}$, specifying the possible time delays of the token before reaching p. In the case of a marked graph the set of feasible delays is described via a set of $(\max, +)$-equations. In [3] it is shown that the design problem for fixed delays can be rewritten as a generalized linear complementarity problem. This can undoubtedly be extended to the case of intervals of delays. These special structures of state machine and marked graph components of $\mathcal{N}_p^*$ will undoubtedly have to be used for the efficient generation and solution of the sets of linear inequalities required for the design of controllers for controlled timed Petri nets.

For general influencing nets the forbidden state feedback control specifies that the disjunction of a number of sets of linear inequalities is satisfied for all possible firing times in the intervals defined by the control being applied. To reduce the complexity of the problem it would be interesting if these inequalities, which have to be satisfied for all possible firing times, could be replaced by conditions on the value of the control function(s) $C_f(M_\theta)$ alone. Note that these inequalities will not only contain the value of the control function containing information on decisions to be made first, but also the set of possible values of a number of future control functions. Control decisions are taken for controllable transitions on the edge of the influencing net as well inside it. An efficient design algorithm will undoubtedly work via decomposing the influencing net in different sub-Petri net, using the topological structure of the net to decompose the conditions on the control functions in a number of simpler conditions, by rearranging the inequalities.

5 References

[1] R.K. Boel, L. Ben-Naoum and V. Van Breusegem, *On Forbidden State Problems for a Class of Controlled Petri Nets*, IEEE-T-AC-40(10), pp. 1717-1731, 1995.

[2] B. Brandin and W. M. Wonham, *Supervisory Control of Timed Discrete Event Systems*, IEEE-T-AC-39(2), pp. 329-342, 1994.

[3] B. De Schutter, *Max-algebraic System Theory for Discrete Event Systems*, Ph.D. Thesis, K.U. Leuven, 1996.

[4] L. Holloway and B.H. Krogh, *Synthesis of Feedback Control Logic for a Class of Controlled Petri Nets*, IEEE-T-AC-35(5), pp. 514-523, 1990.

[5] L. Holloway, *Time Measures and State Maintainability for a Class of Composed Systems*, Proceedings WODES96, pp. 24-30, 1996.

[6] P. Ramadge and W. M. Wonham, *The Control of Discrete Event Systems*, Proceedings IEEE, vol. 77, no.1, pp. 81-98, 1989.

[7] R.K. Boel, B. Bordbar and G. Stremersch, *A Min-Plus Polynomial Approach to Forbidden State Control for General Petri Nets*, Proceedings WODES98, 1998.

14

On matrix mortality in low dimensions

Olivier Bournez* and Michael Branicky**

*VERIMAG
Centre Equation
2, Avenue de Vignate
38610 Gières
FRANCE
bournez@imag.fr

**Electrical Engineering Program
Case Western Reserve University
Cleveland, OH 44106-7221
USA
branicky@alum.mit.edu

1 Description of the problem

A set $F = \{A_1, \dots, A_m\}$ of $n \times n$ matrices is said to be *mortal* if there exist integers $k \geq 1$ and $i_1, i_2, \dots, i_k \in \{1, \dots, m\}$ such that $A_{i_1} A_{i_2} \cdots A_{i_k} = 0$. In that case F is also said to be *k-length mortal.*

We use MORTALITY(n) to denote the class of decision problems "Is a given set F consisting of $n \times n$ matrices mortal?" and MORTALITY(n, m) to denote "Is a given set F of m $n \times n$ matrices mortal?". We also use PAIR-MORTALITY(n) as a synonym for MORTALITY($n, 2$). Unless otherwise noted, all matrices are assumed to have integer-valued entries. But MORTALITY($n, m; \mathbb{R}$), for example, denotes the third problem class for matrices with real-valued entries.

Evidently, MORTALITY(1) and MORTALITY($n, 1$) are efficiently decidable. However, the general complexity of MORTALITY(2) and PAIR-MORTALITY(n), $n < 27$, remains unknown—despite a lot of interest (see [5, 6], which contain some related results, and the references therein).

2 Motivation

Such problems arise as follows:

1. *Controllability of switched linear systems.* Given a system of the form $x(t+1) = A(t,u)x(t)$, where for all t the set of possible values of $A(t,u)$ is a finite set F, the questions above correspond to the controllability (to the origin) of such a system. Cf. [2].

2. MORTALITY(2) is also equivalent to the following problem [8]: Find an algorithm which, given a finite set H of non-singular linear transformations of the complex plane, and lines L and M through the origin, determines whether some product from H maps L onto M.

3 Available results

1. MORTALITY(3) is recursively unsolvable [7]: the proof relies on a reduction of this problem to the Post Correspondence Problem (PCP). It is constructive, using $2p+2$ matrices if PCP is undecidable with p "rules." By considering Modified PCP it is possible to prove undecidability using only $p+2$ matrices [3]. Current bounds on p lie in $\{3, \ldots, 7\}$ (see [1, p. 12] for references and a discussion).

2. Mortality and pair-mortality can be related: if MORTALITY(n,m) is undecidable, then PAIR-MORTALITY(nm) is undecidable [1, 4].

3. PAIR-MORTALITY(2) is decidable [3, 4]. However, the proof uses elementary number theoretic arguments for matrices with complex eigenvalues that do not generalize to matrices with real entries: PAIR-MORTALITY$(2;\mathbb{R})$ has been proved BSS-undecidable [3], yielding

   MORTALITY(n,m) BSS-undecidable for all $n \geq 2$, $m \geq 2$. Nevertheless, PAIR-MORTALITY$(2;\mathbb{R})$ is BSS-decidable for matrices with real eigenvalues [3].

4. PAIR-MORTALITY(n) is decidable and NP-complete when restricted to matrices with non-negative entries [1]. The same argument can be used to show that MORTALITY(n,m) restricted to non-negative matrices is decidable. The problem of deciding whether a given pair of $n \times n$ matrices is k-length mortal, with integer k encoded in unary, is NP-complete; it remains so when the matrices are restricted to have entries in $\{0,1\}$ [1]. The conclusion of NP-completeness in [1] can be more easily obtained using Paterson's construction and reduction to Bounded PCP [3]. The boolean entry case does then not follow, but NP-completeness of "Given a set F of 3×3 matrices and positive integer $K \leq |F|$, is F k-mortal for some $k \leq K$?" does.

4 References

[1] V. D. Blondel and J.N. Tsitsiklis, "When is a pair of matrices mortal?" *Information and Processing Letters*, 63, pp. 283-286 (1997).

[2] V. D. Blondel and J.N. Tsitsiklis, "Complexity of stability and controllability of elementary hybrid systems," Technical Report LIDS-P-2388, Laboratory for Information and Decision Systems, Massachusetts Institute of Technology, Cambridge, MA (1997). Also: *Automatica*, to appear.

[3] O.Bournez and M. S. Branicky, "On the mortality problem for matrices of low dimensions", Technical Report 98-01, VERIMAG, France (1998).

[4] J. Cassaigne and J. Karhumaki, "Examples of undecidable problems for 2-generator matrix semi-groups," Technical Report 57, Turku Center for Computer Science, University of Turku, Turku, Finland (1996).

[5] M. Krom and M. Krom, "Recursive solvability of problems with matrices," *Zwitschr. f. math. Logik und Grundlagen d. Math*, 35, pp. 437–442 (1989).

[6] M. Krom and M. Krom, "More on mortality," *American Mathematical Monthly*, 97, pp. 37–38 (1990).

[7] M. S. Paterson, "Unsolvability in 3 × 3 matrices," *Studies in Applied Mathematics*, XLIX, pp. 105–107 (1970).

[8] P. Schultz "Mortality of 2 × 2 matrices," *American Mathematical Monthly*, 84, pp. 463–464 (1977); correction, 85, p. 263, (1978).

15

Entropy and random feedback

Stephen P. Boyd

Information Systems Laboratory
Electrical Engineering Department
Stanford University
Stanford, CA 94305
USA
boyd@isl.stanford.edu

1 γ-entropy of a matrix

Let $C \in \mathbf{C}^{m \times n}$. The Frobenius norm of C is defined as

$$\|C\|_F = \sqrt{\mathbf{Tr} CC^*},$$

where C^* denotes the complex-conjugate transpose and $\mathbf{Tr}$ denotes trace. The spectral norm of C is defined as

$$\|C\| = \sqrt{\lambda_{\max}(CC^*)},$$

where $\lambda_{\max}$ denotes the maximum eigenvalue.

The problem described in this chapter involves the γ-entropy, which is a convex function of C closely related to these two norms. For $\gamma > 0$ we define the γ-entropy of C as

$$I_\gamma(C) = \begin{cases} -\frac{\gamma^2}{2\pi} \log\det\left(I - \gamma^{-2} CC^*\right) & \text{if } \|C\| < \gamma \\ \infty & \text{if } \|C\| \geq \gamma \end{cases}$$

The Frobenius and spectral norms, and the γ-entropy, are unitarily invariant, and so can be expressed in terms of the singular values σ_i of C (i.e., the squareroots of the eigenvalues of CC^*, ordered as $\sigma_1 \geq \sigma_2 \geq \cdots$):

$$\|C\|_F = \sqrt{\sum_i \sigma_i^2}, \qquad \|C\| = \sigma_1,$$

and

$$I_\gamma(C) = \begin{cases} \frac{1}{2\pi} \sum_i -\gamma^2 \log\left(1 - (\sigma_i/\gamma)^2\right) & \text{if } \sigma_1 < \gamma \\ \infty & \text{if } \sigma_1 \geq \gamma \end{cases}.$$

The following facts are readily shown (see [1, §5.3.5]). The squareroot of the γ-entropy always exceeds the Frobenius norm:

$$\sqrt{I_\gamma(C)} \geq \|C\|_F,$$

and as γ becomes large, it converges to the Frobenius norm:

$$\lim_{\gamma\to\infty} \sqrt{I_\gamma(C)} = \|C\|_F.$$

We also have the more complicated converse inequality

$$\sqrt{I_\gamma(C)} \leq \frac{1}{\alpha}\sqrt{-\log(1-\alpha^2)}\,\|C\|_F,$$

where $\alpha = \|C\|/\gamma < 1$. Thus, the relative increase in the squareroot of the γ-entropy over the Frobenius norm can be bounded by an expression that only depends on how close the spectral norm is to the critical value γ. For example, if $\|C\| \leq \gamma/2$ ($\alpha \leq 0.5$), we have $\|C\|_F \leq \sqrt{I_\gamma(C)} \leq 1.073\|C\|_F$. The γ-entropy arises in several contexts, for example $\mathbf{H}_\infty$ control, in which the so-called central controller minimizes the integral over frequency of the γ-entropy of the closed-loop transfer matrix (see, e.g., [2, 1]). It arises as the natural self-concordant barrier for the convex set $\{ C \mid \|C\| \leq \gamma \}$, in interior-point optimization methods (see [4, 3]). The γ-entropy also arises in other applications, e.g., contractive matrix completion problems [5].

2 Stochastic interpretation

The Frobenius norm can be interpreted as the root-mean-square gain of the matrix C, as follows. Suppose $w \in \mathbf{C}^n$ is a random vector with zero mean and covariance I, i.e.,

$$\mathbf{E}w = 0, \qquad \mathbf{E}ww^T = I,$$

and let $z = Cw$. Then we have

$$\mathbf{E}\|z\|^2 = \|C\|_F^2.$$

We now connect a feedback gain $\Delta \in \mathbf{C}^{n\times m}$ around C, i.e., we consider

$$z = Cu, \qquad u = w + \Delta z.$$

Eliminating u yields the familiar formula for the 'closed-loop gain':

$$z = C(I - \Delta C)^{-1}w.$$

(The inverse exists if the 'small-gain' condition $\|\Delta\| < \gamma$ holds.) Evidently the root-mean-square value of z is given by $\|C(I-\Delta C)^{-1}\|_F$.

Now we assume that Δ is a random matrix, independent of w, such that $\|\Delta\| < \gamma$ with probability one. The mean-square value of z is then

$$\mathbf{E}\|z\|^2 = \mathbf{E}\|C(I - \Delta C)^{-1}\|_F^2.$$

where the expectation is over the random feedback gain Δ.
Our open problem can now be stated:

> *Find a distribution for Δ (if one exists) such that the mean-square value of z is given by $I_\gamma(C)$, i.e., $I_\gamma(C) = \mathbf{E}\|C(I - \Delta C)^{-1}\|_F^2$.*

Evidently the distribution should be unitarily invariant, and must satisfy $\|\Delta\| < \gamma$ with probability one.
If such a distribution can be found we will have a nice interpretation of the entropy as the mean-square value of the output of a system, with a random input and a random feedback connected around it. The inequalities above would then show that the random feedback has little effect unless the norm of the feedback is significant compared to the norm of C.

3 The scalar case

The problem has been solved for the scalar case $m = n = 1$ in [1]. If Δ is uniformly distributed on the disk of radius $1/\gamma$ in the complex plane, then we have

$$I_\gamma(C) = \mathbf{E}\|C(I - \Delta C)^{-1}\|_F^2 = \mathbf{E}\left|\frac{C}{1 - \Delta C}\right|^2.$$

This can be shown as follows.

$$\begin{aligned}\mathbf{E}\left|\frac{C}{1 - \Delta C}\right|^2 &= \frac{\gamma^2}{\pi}\int_0^{1/\gamma}\int_0^{2\pi}\left|\frac{C}{1 - re^{i\theta}C}\right|^2 r\,d\theta\,dr \\ &= \begin{cases} -\gamma^2\log(1 - |C|^2/\gamma^2) & |C| < \gamma \\ \infty & |C| \geq \gamma \end{cases}\end{aligned}$$

(the integration over θ can be evaluated by residues).

4 References

[1] S. Boyd and C. Barratt. *Linear Controller Design: Limits of Performance.* Prentice-Hall, 1991.

[2] D. Mustafa and K. Glover. *Minimum Entropy* $\mathbf{H}_\infty$ *Control.* Lecture Notes in Control and Information Sciences. Springer-Verlag, 1990.

[3] Yu. Nesterov and A. Nemirovsky. *Interior-point polynomial methods in convex programming*, volume 13 of *Studies in Applied Mathematics*. SIAM, Philadelphia, PA, 1994.

[4] L. Vandenberghe and S. Boyd. Semidefinite programming. *SIAM Review*, 38(1):49–95, March 1996.

[5] L. Vandenberghe, S. Boyd, and S.-P. Wu. Determinant maximization with linear matrix inequality constraints. *SIAM J. on Matrix Analysis and Applications*, April 1998. To appear.

16

A stabilization problem

Roger Brockett

Division of Engineering and Applied Sciences
Harvard University
USA
brockett@metatron.harvard.edu

1 Background

It is, of course, well known that the usual output stabilization problem, asking about the existence of a constant feedback gain that stabilizes the system, does not have a especially clean answer. Here we describe an alternative version involving the existence of time-varying gains. The problem is this.

Problem 1: Given a triple of constant matrices (A, B, C) under what circumstances does there exist a time dependent matrix K such that the system

$$\dot{x}(t) = Ax(t) + BK(t)Cx(t)$$

is asymptotically stable.

To avoid tediousness, assume that rank $(B, AB, ..., A^{n-1}B) = n$ and rank $(C, CA, ..., CA^{n-1}) = n$ where n is the dimension of x. It seems that little is lost by assuming that there exists a time T such that K is periodic of period T.

We make a few remarks about what is already known.

1. Suppose that K is m by p so that its entries can be thought of varying over a mp-dimensional vector space. In the case $mp \geq n$ a dimension counting argument would suggest that the problem should have a positive answer, at least for large classes of A, B, and C. One might even expect to be able to position the entire set of eigenvalues of $A + BKC$ and Rosenthal has proved that for generic (A, B, C) such a result holds over the complex field [2].

2. One way to think about this problem is to consider the entries of K to be controls and to rephrase the problem as a controllability problem.

Problem 1': Given a triple of constant matrices (A, B, C) under what circumstances does there exist a time dependent control K such that the system

$$\dot{X}(t) = (A + BK(t)C)X(t) \ ; \ X(0) = I$$

has the matrix 0 in the closure of its reachable set.
Of course the controllability problem immediately suggests that the Lie algebra generated by the set of all matrices of the form $A + BKC$, for all values of K, should play a role. See [1]. We will return to an aspect of that below but first we observe the following necessary condition.

Necessary Condition 1: It is necessary that for each initial condition $x(0)$ there should exist a control K defined on $[0, \infty)$ such that the solution of the vector equation

$$\dot{x}(t) = Ax(t) + BK(t)Cx(t)$$

approaches zero as t goes to infinity.
This condition is not sufficient. It may happen that it is possible to drive each initial condition to zero and yet not be able to find a suitable "universal" K. This phenomenon is partially explained by the following easily proved necessary condition.

Necessary Condition 2: It is necessary that either tr$A$ is negative or else that CB is non zero.
To prove this we appeal to the Liouville identity

$$\det(X)(t) = e^{\int_0^t \mathrm{tr}(A+BK(\sigma)C)d\sigma} \det X(0)$$

and observe that if we can not force

$$\mathrm{tr}(A + BK(t)C) = tr(A) + \mathrm{tr}(KCB)$$

to be negative we can not stabilize the system. This condition can be sharpened to the following extent. Consider the linear span of all matrices of the form $A + BKC$ as K ranges over all possible m by p matrices. Denote this by $\mathcal{L}$. If there exists an invertible matrix P such that $P\mathcal{L}P^{-1}$ is block diagonal then the trace condition must hold for each block individually. We can call this the *reducible* case. Without loss of generality we can restrict discussion to the irreducible case.

2 When K is a scalar

The problem is still interesting and unsolved in the case where K is a scalar. Of course even in the original situation the existence of a stabilizing

K depends only on the rational function $G(s) = C(Is - A)^{-1}B$ because it incorporates all "basis free" aspects of the problem. If K is a scalar then so is this rational function and we can ask for an answer to our original problem in terms of it. In this case there is a quite specific test that goes beyond any of the necessary conditions mentioned above. It defines a third necessary condition.

Necessary Condition for Scalar K: It is impossible to find a time-varying gain that stabilizes the given system if there exists an invertible matrix P such that

$$P(A + BKC)P^{-1} = \begin{bmatrix} A_{11} & A_{12} \\ A_{21} & \bar{k} \end{bmatrix}$$

with

$$PAP^{-1} = \begin{bmatrix} A_{11} & A_{12} \\ A_{21} & 0 \end{bmatrix}$$

being nonnegative off the diagonal and A_{11} being unstable.

To prove this we recall that any solution of $\dot{x} = A(t)x$ that starts with all components of x nonnegative will continue with all elements of x nonnegative if A is nonnegative off the diagonal. In this situation x will stay nonnegative regardless of the choice of k. If we denote the first $n-1$ components of x by x_u then of course

$$\dot{x}_u = A_{11}x_u + A_{12}x_n$$

or

$$x_u(t) = e^{A_{11}t}x_u(0) + \int_0^t e^{A_{11}(t-\sigma)}A_{12}x_n(\sigma)d\sigma.$$

However, we only make x_u smaller by eliminating the integral term and $e^{A_{11}t}x_u(0)$ is unbounded except for possibly a codimension one or higher subset of the initial conditions.

As an example, this can be used to show that if the poles and zeros of $g(s)$ are all real and interlacing, and if at least two of the poles lie in the right half-plane then the system can not be stabilized by constant output feedback.

3 A Generalization

A more general problem along these lines is this.

Problem 1": Given a family of constant matrices

$$(A, B_1, B_2, ..., B_r, C_1, C_2, ..., C_r)$$

under what circumstances do there exist time dependent matrices K_1, K_2, ..., K_r such that the system

$$\dot{x}(t) = Ax(t) + \sum B_i K_i(t) C_i x(t)$$

is asymptotically stable.
To illustrate one of the subtleties associated with this problem we cite an example that arose in the study of a related nonlinear problem.

Remark: It is possible to find periodically varying functions $k_1(t)$ and $k_2(t)$ such that

$$\dot{X}(t) = \begin{bmatrix} -1 & 0 & k_1(t) \\ 0 & -1 & k_2(t) \\ k_2(t) & -k_1(t) & 0 \end{bmatrix} X(t)$$

is asymptotically stable.

4 References

[1] R. W. Brockett, "System Theory on Group Manifolds and Coset Spaces," *SIAM Journal on Control Optim.*, Vol. 10 (1972) pp. 265-284.

[2] J. Rosenthal, "On dynamic feedback compensation and compactification of systems," *SIAM Journal on Control Optim.*, Vol. 32 (1994) pp. 279-296.

17
Spectral factorization of a spectral density with arbitrary delays

Frank M. Callier and Joseph J. Winkin

Department of Mathematics
Facultés Universitaires Notre-Dame de la Paix
Rempart de la Vierge 8
5000 Namur
BELGIUM
Frank.Callier@fundp.ac.be
Joseph.Winkin@fundp.ac.be

1 Description of the problem

An impulse response f (with support on the nonnegative real numbers) is said to be in $\mathcal{A}_-$ if, on $t \geq 0, f(t) = f_a(t) + f_{sa}(t)$ where its regular functional part f_a is such that $\int_0^\infty |f(t)| \exp(-\sigma t) dt < \infty$ and its singular atomic part $f_{sa} := \sum_{i=0}^{\infty} f_i \delta(. - t_i)$, where $t_0 = 0, t_i > 0$ for $i \geq 1$ and $f_i \in \mathbb{C}$ for $i \geq 0$ with $\sum_{i=0}^{\infty} |f_i| \exp(-\sigma t_i) < \infty$, for some $\sigma < 0$. $\hat{\mathcal{A}}_-$ denotes the algebra of distributed parameter system proper-stable transfer functions, i.e. Laplace transforms of elements in $\mathcal{A}_-$.
Let $F = F_a + F_{sa} := F_a(\cdot) + \sum_{i=-\infty}^{\infty} F_i \delta(\cdot - t_i)$ be a matrix of Laplace transformable distributions of a real variable $t \in \mathbb{R}$, where a) $F_a(.)$ is a $n \times n$-matrix valued function, F_i is a constant $n \times n$-matrix, for $i = 0, \pm 1, \cdots$, $t_0 = 0$, $t_i > 0$ and $t_{-i} < 0$ for $i \geq 1$, and $\delta(\cdot)$ denotes the Dirac delta distribution (Dirac impulse); b) F is parahermitian self-adjoint, i.e. $F(t) = F_*(t) := F(-t)^*$, i.e. $F_a(t) = (F_a)_*(t)$, $F_{-i} = F_i^*$ and $t_{-i} = -t_i$ for $i \geq 0$; c) the Laplace transform of its causal part (i.e. its part with support on the nonnegative real numbers), viz. $F^+ = F_a^+(.) + 2^{-1} F_0 \delta(.) + \sum_{i=1}^{\infty} F_i \delta(. - t_i)$, is a matrix with all its entries in the algebra $\hat{\mathcal{A}}_-$; and d) $\hat{F}$ is positive semi-definite on the imaginary axis. We are interested in finding a necessary and sufficient condition on the matrix spectral density $\hat{F}$ under which it admits a *spectral factor*, i.e. an invertible matrix valued function $\hat{R} = \hat{R}_a + \hat{R}_{sa} = \hat{R}_a(\cdot) + \sum_{k=0}^{\infty} R_k \cdot \exp(- \cdot \ \tau_k)$ with all entries in $\hat{\mathcal{A}}_-$ together with its inverse, such that $\hat{F}(j\omega) = \hat{R}(j\omega)^* \cdot \hat{R}(j\omega)$ for all $\omega \in \mathbb{R}$.

2 Motivations

This problem is encountered, under different forms, in several kinds of applications.

1. A classical application is the solution of Wiener-Hopf type problems, i.e. *systems of integral equations* on a half line, with kernels of a specific type. A lot of results in this field have been developed by Gohberg and Krein: see e.g. [6] - [7] and the book [5].

2. Another context in which such a problem arises is the theory of feedback control system design and more specifically the multiplier technique in passivity theory: see e.g. [10], [4].

3. The specific spectral factorization problem described above is motivated by applications in feedback control system design, more precisely in the analysis of *closed-loop stability robustness*, i.e. the analysis of the graph distance between two possibly unstable systems for obtaining robustness estimates of feedback stability, and in the solution to the *Linear-Quadratic optimal control* problem by frequency domain techniques for distributed parameter, i.e. infinite-dimensional state-space, systems, in the context of the transfer function algebra developed by Callier and Desoer: see e.g. [1], [11], [2] and [9] and references therein.

3 Available results

An easy necessary condition is the following inequality: $inf\{det\hat{F}(j\omega) : \omega \in \mathbb{R}\} > 0$, which is known to be necessary and sufficient in the H_∞ setting, see [8, p. 54; Theorem 6.14, p. 124]. It turns out that this condition is also sufficient in the $\hat{\mathcal{A}}_-$ setting, in several particular situations:

1. The problem is solved for the monovariable case of a spectral density giving a spectral factor invertible in $\hat{\mathcal{A}}_-$ with delayed impulses: see [1], where the analysis is mainly based on [5, pp. 158-160].

2. In the multivariable case, the problem is solved in [2] for a matrix spectral density with no delayed impulses: its solution is essentially based on [6, Theorem 8.2] and an analytic extension technique similar to that in [7, Proof of Theorem 3.1].

3. The mutivariable case giving a matrix spectral factor with all entries in $\hat{\mathcal{A}}_-$ together with its inverse, with commensurate, i.e. equally spaced, delays (e.g. transmission lines) is solved in [11],[3]. In this

case, the main step is a conformal mapping transformation which reduces the problem to a factorization problem on the unit circle, see [6, §14].

4. The multivariable case of a matrix spectral density with uncommensurate, i.e. unequally spaced, delays is an open question: see [5, p. 252, Comments on Chapter VIII]. Only a sufficient condition is known, which is stronger than the condition given above and under which the problem reduces to a fixed point equation leading to a causal power series expansion of the spectral factor: see [3], where the analysis is based on [4, Section 9.5] and [10, Section 3.7].

4 References

[1] F.M. Callier and J. Winkin, "The spectral factorization problem for SISO distributed systems", in *Modelling, robustness and sensitivity reduction in control systems* (R.F. Curtain, ed.), NATO ASI Series, Vol. F34, Springer-Verlag, Berlin Heidelberg, pp. 463–489 (1987).

[2] F.M. Callier and J. Winkin, "On spectral factorization and LQ-optimal regulation for multivariable distributed systems", *International Journal of Control*, 52, No. 1, pp. 55-75 (1990).

[3] F.M. Callier and J. Winkin, "The spectral factorization problem for multivariable distributed parameter systems", FUNDP Namur, Department of Mathematics, Report 97-06 (1997); *Integral Equations ans Operator Theory*, submitted.

[4] C.A. Desoer and M. Vidyasagar, "Feedback systems: Input-Output properties", Academic Press, New York (1975).

[5] I.C. Gohberg and I.A. Fel'dman, "Convolution equations and projection methods for their solution", AMS Translations, Providence, RI (1974).

[6] I.C. Gohberg and M.G. Krein, "Systems of integral equations on a half line with kernels depending on the difference of arguments", AMS Translations, Vol. 2, pp. 217–287 (1960).

[7] M.G. Krein, "Integral equations on a half line with kernels depending upon the difference of the arguments", AMS Translations, Vol. 22, pp. 163–288 (1962).

[8] M. Rosenblum and J. Rovnyak, "Hardy classes and operator theory", Oxford University Press, New York (1985).

[9] O.J. Staffans, "Quadratic optimal control through coprime and spectral factorizations", rAbo Akademi Reports on Computer Science and Mathematics, Ser. A, No. 178 (1996).

[10] J.C. Willems, "The analysis of feedback systems", The M.I.T. Press, Cambridge, MA (1971).

[11] J. Winkin, "Spectral factorization and feedback control for infinite-dimensional systems", Doctoral Thesis, Department of Mathematics, Facultés Universitaires Notre-Dame de la Paix, Namur, Belgium (1989).

18

Lyapunov exponents and robust stabilization

F. Colonius*, D. Hinrichsen** and F. Wirth**

*Institut für Mathematik
Universität Augsburg
86135 Augsburg
GERMANY
colonius@math.uni-augsburg.de

**Institut für Dynamische Systeme
Universität Bremen
28334 Bremen
GERMANY
dh@mathematik.uni-bremen.de
fabian@math.uni-bremen.de

1 Description of the problems

In recent years there has been considerable interest in the analysis and synthesis of time-varying control systems. Here we present some open problems in this area. Let $\mathbf{K} = \mathbb{R}, \mathbb{C}$ and consider a family of bilinear systems

$$\begin{aligned} \dot{x}(t) &= \left(A_0 + \sum_{i=1}^{m} v_i(t) A_i \right) x(t) \qquad (18.1) \\ x(0) &= x^0 \in \mathbf{K}^n \\ v(t) &= (v_1(t), \ldots, v_m(t))' \in V \text{ a.a. } t \geq 0 \end{aligned}$$

where $A_0, \ldots, A_m \in \mathbf{K}^{n \times n}$ and $V \subset \mathbf{K}^m$ is a convex, compact set with $0 \in \operatorname{int} V$. The set of admissible control functions $\mathcal{V}$ is the set of measurable functions $v : \mathbb{R} \to V$. The solution determined by an initial condition x^0 and a control $v \in \mathcal{V}$ is denoted by $\varphi(\cdot; x^0, v)$.
The exponential growth rate or *Lyapunov exponent* of a trajectory with initial state $0 \neq x^0 \in \mathbf{K}^n$ and control $v \in \mathcal{V}$ is defined by

$$\lambda(x^0, v) := \limsup_{t \to \infty} \frac{1}{t} \log \|\varphi(t; x^0, v)\| .$$

It is well known that for fixed $v \in \mathcal{V}$ there are at most n different Lyapunov exponents of (18.1). For constant controls $v(\cdot) \equiv v$ the Lyapunov exponents are simply the real parts of the eigenvalues of $A_0 + \sum_{i=1}^{m} v_i A_i$. If $v \in \mathcal{V}$ is periodic the Lyapunov exponents given by v are the *Floquet exponents* from classical Floquet theory. The Lyapunov and the Floquet spectrum are defined by

$$\Sigma_{Ly} := \{\lambda(x^0, v)\ ;\ 0 \neq x^0 \in \mathbf{K}^n,\ v \in \mathcal{V}\},$$

$$\Sigma_{Fl} := \{\lambda(x^0, v)\ ;\ 0 \neq x^0 \in \mathbf{K}^n,\ v \in \mathcal{V}, v \text{ periodic}\}.$$

Under suitable assumptions the closure of the Floquet spectrum $\mathrm{cl}\,\Sigma_{Fl}$ is given by intervals whose number does not exceed the dimension of the state space n, see [1] for the continuous and [8] for the discrete-time case. The investigation of this property is closely related to controllability properties of the system obtained by projection onto the sphere. In general these spectral intervals are only accessible through involved numerical computations except for easy cases like the following example taken from [1].
Consider the system

$$\dot{x}(t) = \left(\begin{bmatrix} 0 & 1 \\ 2 & 0 \end{bmatrix} + v(t) \begin{bmatrix} 0 & 0 \\ 0 & 1 \end{bmatrix} \right) x(t).$$

For a control range $V = [-\rho, \rho]$ it may be shown that the Floquet intervals are given by

$$\left(-\frac{\rho}{2} - \sqrt{2 + \frac{\rho^2}{4}}, \frac{\rho}{2} - \sqrt{2 + \frac{\rho^2}{4}} \right), \quad \left(-\frac{\rho}{2} + \sqrt{2 + \frac{\rho^2}{4}}, \frac{\rho}{2} + \sqrt{2 + \frac{\rho^2}{4}} \right).$$

Furthermore, in this example the Lyapunov spectrum is equal to the closure of the Floquet spectrum. It is still not known whether this is always the case.

Problem 1: Floquet versus Lyapunov spectrum. Prove or disprove that for all systems of the form (18.1)

$$\Sigma_{Ly} = \mathrm{cl}\,\Sigma_{Fl}.$$

In the formulation of (18.1) the v_i may be interpreted as time-varying perturbations or as open-loop controls. We will now concentrate on the interpretation of v as a time-varying disturbance counteracting a linear control input into the system. Thus we consider the following system

$$\dot{x}(t) = \left(A_0 + \sum_{i=1}^{m} v_i(t) A_i \right) x(t) + Bu(t), \tag{18.2}$$

where the assumptions on the disturbance v and the set V are as before and $B \in \mathbf{K}^{n \times p}, u(t) \in \mathbf{K}^p$.

We propose to consider the following robust stabilization problem: Construct a (possibly time-varying) feedback $F : \mathbb{R}_+ \to \mathbf{K}^{p\times n}$ such that

$$\dot{x}(t) = \left(A_0 + \sum_{i=1}^{m} v_i(t)A_i\right) x(t) + BF(t)x(t), \qquad (18.3)$$

is exponentially stable for all disturbances $v \in \mathcal{V}$. Such a map F is called *robustly stabilizing state-feedback.* A further question of interest is to restrict the set of admissible values for F, namely to require that $F(t) \in \Gamma \subset \mathbf{K}^{p\times n}$, where Γ satisfies a set of conditions similar to those on V.

Problem 2: Robust stabilization. Develop a (game theoretic?) robust stabilization approach for system (18.2). In particular, solve the following problems.

Given the matrix A_0, the perturbation structure $A_1, \ldots, A_m$, the perturbation set $\mathcal{V}$ and the gain restriction Γ.

a) Find necessary and sufficient conditions for the existence of an admissible robustly stabilizing time-invariant state-feedback.

b) Find necessary and sufficient conditions for the existence of an admissible robustly stabilizing periodic state-feedback.

c) Determine conditions under which the existence of a time-varying admissible robustly stabilizing feedback is equivalent to the existence of an admissible time-invariant (periodic) one.

d) Develop algorithms for the computation of a robustly stabilizing feedback controllers for (18.2).

2 History of the problem

Exponential growth rates of time-varying linear systems have already been studied by Floquet and Lyapunov. A further breakthrough was achieved in 1968 when Oseledets proved the famous multiplicative ergodic theorem which showed under ergodicity assumptions the existence of well behaved Lyapunov exponents in the stochastic framework. This result motivated a flurry of further research on Lyapunov exponents in smooth ergodic theory and for stochastic systems, which in turn motivated a similar analysis in the deterministic case over the last decade.

3 Motivations

For a large class of systems where Floquet and Lyapunov spectra coincide approximations to the spectra can be computed by optimal control methods (via numerical solutions of Hamilton-Jacobi-Bellman equations, see [3]), thus a positive solution of Problem 1 also shows computability of the Lyapunov spectrum. This is of particular interest as it has been shown in [6] that there is no algorithm to compute Lyapunov exponents in the stochastic framework in the general case. The complete Lyapunov spectrum can be used for different problem areas arising in control:

1. Robustness analysis: Interpreting v as a disturbance, the time-varying stability radius of a stable matrix A_0 corresponding to perturbations of the form (18.1) can be defined as the infimum of those values $\rho \in \mathbb{R}_+$ for which the maximal Lyapunov exponent of (18.1) with control range $V_\rho := \rho V$ is nonnegative.

2. Stabilization: Interpreting v as a control input, the existence of an appropriately defined feedback that stabilizes (18.1) can be characterized in terms of properties of the Lyapunov spectrum. In contrast to the first item knowledge of the complete spectrum is necessary.

Problem 2 is similar to the robust stabilization problems that have been at the center of control theory in recent years. The following singular H_∞ problem is a special case (with *constant unconstrained* feedbacks F): Given $A \in \mathbf{K}^{n\times n}$, $D \in \mathbf{K}^{n\times \ell}$, $E \in \mathbf{K}^{q\times n}$ and $\rho > 0$, find a feedback matrix $F \in \mathbf{K}^{p\times n}$ such that all the systems

$$\dot{x}(t) = (A + D\Delta(t)E + BF)x, \quad \Delta(\cdot) \in L_\infty(0,\infty;\mathbf{K}^{\ell\times q}),\ \|\Delta(\cdot)\|_{L_\infty} \leq \rho \tag{18.4}$$

are stable. This problem may be formulated as in Problem 2 by choosing a basis $A_1, \dots, A_m$ of the subspace $\{D\Delta E \in \mathbf{K}^{n\times n}\ ;\ \Delta \in \mathbf{K}^{\ell\times q}\}$ and considering the system

$$\dot{x}(t) = (A + \sum_{i=1}^{m} v_i(t)A_i + BF)x\,,$$

where $V = \{v \in \mathbf{K}^m\ ;\ \exists\Delta : \|\Delta\| \leq \rho \text{ and } \sum_{i=1}^m v_iA_i = D\Delta E\}$.
Note, however, that Problem 2 encompasses a far larger problem class. For instance, feedback constraints are directly taken into account. Furthermore, arbitrary time-varying affine parameter perturbations can be represented as in (18.2) so that Problem 2 has features of a time-varying μ-synthesis problem.

4 Available results

The Lyapunov spectrum of families of bilinear systems of the form (18.1) has been studied intensively, e.g. in [1], [2]. In these publications the bilinear system is projected onto the projective space $\mathbf{PS}^{n-1}$ and controllability properties of the projected nonlinear system are used to characterize the Floquet spectrum. The main assumption used in this approach is that the

nonlinear system is locally accessible. Then by a result due to E. Joseph, the equality in Problem 1 holds in dimension n=2. Furthermore, it has been shown in the discrete-time case that $\mathrm{cl}\,\Sigma_{Fl} \subset \Sigma_{Ly}$ using a method that naturally extends to the continuous-time case, [8]. This result is based on the standard accessibility assumptions which allow in particular to construct Lyapunov exponents from sequences of Floquet exponents by connecting periodic orbits.
Considering scaled ranges $V_\rho := \rho V$ it is shown in [1] that, under an additional controllability condition, $\mathrm{cl}\,\Sigma_{Fl}(\rho) = \Sigma_{Ly}(\rho)$ for all but countably many $\rho > 0$. Here $\Sigma_{Ly}(\rho)$ and $\Sigma_{Fl}(\rho)$ denote the Lyapunov and Floquet spectra of (18.1) with perturbation range restricted to V_ρ.
So far these results have been applied to the stabilization problem [3], [7]. A numerical stabilization scheme based on the theory of Lyapunov exponents is described in [3]. Robustness analysis has been treated in [9].
Problem 2 has hardly been dealt with. In the complex case $\mathbf{K} = \mathbb{C}$ the singular H_∞ problem (18.4) is solvable if and only if there exists $F \in \mathbb{C}^{m\times n}$ such that the complex stability radius $r_{\mathbb{C}}(A+BF, D, E) > \rho$, see [4]. Necessary and sufficient conditions for the existence of such a feedback are given in [5]. In the real case $\mathbf{K} = \mathbb{R}$ the problem is still unsolved.

5 References

[1] F. Colonius and W. Kliemann. The Lyapunov spectrum of families of time varying matrices. *Trans. Am. Math. Soc.*, 348:4389–4408, 1996.

[2] F. Colonius and W. Kliemann. The Morse spectrum of linear flows on vector bundles. *Trans. Am. Math. Soc.*, 348:4355–4388, 1996.

[3] L. Grüne. Discrete feedback stabilization of semilinear control systems, ESAIM: Control, Optimisation and Calculus of Variations, 1:207-224, 1996.

[4] D. Hinrichsen and A. J. Pritchard. Real and complex stability radii: a survey. In D. Hinrichsen and B. Mrartensson, editors, *Control of Uncertain Systems*, volume 6 of *Progress in System and Control Theory*, pages 119–162, Basel, 1990. Birkhäuser.

[5] D. Hinrichsen, A. J. Pritchard, and S. B. Townley. A Riccati equation approach to maximizing the complex stability radius by state feedback. *Int. J. Control*, 52:769–794, 1990.

[6] J. Tsitsiklis and V. Blondel. The Lyapunov exponent and joint spectral radius of pairs of matrices are hard - when not impossible - to compute and to approximate. *Math. Control Signals Syst.*, 10:31-40, 1997.

[7] H. Wang, Feedback stabilization of bilinear control systems. PhD thesis, Iowa State University, Ames, IA, 1998.

[8] F. Wirth. Dynamics of time-varying discrete-time linear systems: Spectral theory and the projected system. *SIAM J. Contr. & Opt.*, 36(2):447-487, 1998.

[9] F. Wirth. On the calculation of real time-varying stability radii. *Int. J. Robust & Nonlinear Control*, 1998. To appear.

19

Regular spectral factorizations

Ruth F. Curtain* and Olof J. Staffans**

*Department of Mathematics
University of Groningen
P.O. Box 800
9700 AV Groningen
THE NETHERLANDS
R.F.Curtain@math.rug.nl

**Abo Akademi University
Department of Mathematics
20500 Abo
FINLAND
Olof.Staffans@abo.fi

1 Description of the Problem

Problem 1 *Let $\mathcal{X}$ be an an invertible $n \times n$ matrix-valued function in the open right half plane, which belongs to H^∞ together with its inverse. Find conditions on the boundary function $\Pi(i\omega) = \mathcal{X}(i\omega)^*\mathcal{X}(i\omega)$ ($\omega \in \mathbb{R}$) which imply that the limit $E = \lim_{\lambda \to +\infty} \mathcal{X}(\lambda)$ exists, and compute this limit. Here λ tends to infinity along the positive real axis.*

By mapping the right half-plane into the unit disk we obtain the following equivalent formulation of our problem:

Problem 2 *Let $\mathcal{X}$ be an an invertible $n \times n$ matrix-valued function in the open unit disk, which belongs to H^∞ together with its inverse. Find conditions on the boundary function $\Pi(z) = \mathcal{X}(z)^*\mathcal{X}(z)$ ($|z| = 1$) which imply that the limit $E = \lim_{\eta \to 1-} \mathcal{X}(\eta)$ exists, and compute this limit. Here η tends to 1 from the left along the positive real axis.*

The function $\mathcal{X}$ is called a *spectral factor* of the "Popov function" Π, and this spectral factor is *regular* if the limit E exists. Observe that $\Pi(i\omega) = |X(i\omega)|^2$ in the important scalar case. There is also an infinite-dimensional version of the same problem, where the values of $\mathcal{X}$ lie in the space of bounded linear operators on a separable Hilbert space U. In this case, we obtain three different versions of the problem, depending on the topology we use to compute the limit E, i.e, as a uniform operator limit, or as a strong limit, or as a weak limit. In this case we would also like to know if E is invertible.

2 History of the Problem and Motivations

This open problem arose fairly recently out of some new results on infinite-dimensional regulator problems and the corresponding Riccati equations for the very general class of infinite-dimensional linear system known as (weakly) *regular well-posed linear systems*, see [5, 8, 9, 10, 14, 15, 17]. The existence of solutions to Riccati equations is the key to solving many control problems and it is important to prove existence for as large a class of systems as possible. The approach used here is to reduce the regulator problem to the associated spectral factorization problem, appeal to known results for the existence of a spectral factor $\mathcal{X}$, and then to use the beautiful properties of well-posed linear systems to construct a realization of this spectral factor from a given realization of the original system. If the spectral factor is regular, the appropriate algebraic Riccati equations can be derived and the optimal solution can be constructed. The conclusions are perfect generalizations of the corresponding finite-dimensional conclusions (see [3]), apart from the following gaps:

- Although, under the standard assumptions, the existence of a spectral factor is guaranteed, it need not be regular, i.e., the limit E need not exist. This regularity property is essential in the derivation of the Riccati equation.
- The resulting Riccati equation contains the operator $(E^*E)^{-1}$. Hence we need to compute E^*E, and E^*E must be invertible.

3 Known Results

Different sets of conditions are known that imply that the spectral factor is regular and that the limit E can be computed. They are all in one way or another related to the smoothness of the Popov function Π, which in terms of the original system typically can be written in the form

$$\Pi(i\omega) = R + NG(i\omega) + (NG(i\omega))^* + G(i\omega)^*QG(i\omega),$$

where $G(s) = C(sI - A)^{-1}B$ is the system transfer function and R, Q, and N are various weighting operators. The corresponding Riccati equation Π and the equation for the optimal feedback operator K are in this case

$$A^*\Pi + \Pi A + Q \quad = \quad (B^*\Pi + N)^*(E^*E)^{-1}(B^*\Pi + N),$$

$$K = -(E^*E)^{-1}(B^*\Pi + N).$$

We have (partially) positive answers in the following cases (here we assume for simplicity that A generates an exponentially stable semigroup; see [1, 4, 6, 13, 17] for more details):

- The original system has a bounded control operator B and an admissible observation operator C, e.g., it is of Pritchard-Salamon type;
- The dimension of the output space is finite, and the original system has an admissible control operator B and a bounded observation operator C;
- A generates an exponentially stable analytic semigroup, and $C(-A)^{-\gamma}B$ is bounded for some $\gamma < 1$;
- The dimension of the input and output spaces are finite, and the impulse response of the original system is an L^1-funtion;
- The dimensions of the input and output spaces are finite, and $G(s)$ is a rational function of e^{-Ts}, for some $T0$.

Some combinations of these cases are also possible. It is known that the spectral factor need not be regular; for an example see [17]. Some other illustrating examples are given in [7, 16]. What is lacking are verifiable conditions under which the spectral factor is regular. In particular, it is not known to what extent systems modelled as a boundary control problem for the wave equation in several space dimensions have a regular spectral factor. An abstract formula for the limit E is given in [7, 8] and, as one would expect, in the classical case and all the first 4 special cases itemized above the limit satisfies $E^*E = R$. What we would like to be able to do is to compute it in all cases.

4 Related Results

The same problem appears in a more general setting where one looks for sign-indefinite J-spectral factors of a given boundary function, and derives the corresponding algebraic Riccati equations. These become important in the solution of H^∞ type problems; see [4, 11, 12]. In another generalization we keep the assumption that $\mathcal{X}$ is invertible in the open right half plane and "outer", but we do not require the inverse to be in H^∞. This case is related to the so called singular regulator problem where $R = 0$, and clearly we expect E to be zero. Some sufficient conditions for this to be true are given in [2].

5 References

[1] Frank M. Callier and Joseph Winkin. LQ-optimal control of infinite-dimensional systems by spectral factorization. *Automatica*, 28:757–770, 1992.

[2] Ruth F. Curtain. Linear operator inequalities for stable weakly regular linear systems. Preprint, 1997.

[3] Vlad Ionescu and Martin Weiss. Continuous and discrete-time Riccati theory: a Popov-function approach. *Linear Algebra and its Applications*, 193:173–209, 1993.

[4] Kalle Mikkola. On the stable H^2 and H^∞ infinite-dimensional regulator problems and their algebraic Riccati equations. Technical Report A383, Institute of Mathematics, Helsinki University of Technology, Espoo, Finland, 1997.

[5] Dietmar Salamon. Realization theory in Hilbert space. *Mathematical Systems Theory*, 21:147–164, 1989.

[6] Olof J. Staffans. Quadratic optimal control of stable systems through spectral factorization. *Mathematics of Control, Signals, and Systems*, 8:167–197, 1995.

[7] Olof J. Staffans. On the discrete and continuous time infinite-dimensional algebraic Riccati equations. *Systems and Control Letters*, 29:131–138, 1996.

[8] Olof J. Staffans. Quadratic optimal control of stable well-posed linear systems. To appear in Transactions of American Mathematical Society, 1997.

[9] Olof J. Staffans. Coprime factorizations and well-posed linear systems. To appear in SIAM Journal on Control and Optimization, 1997.

[10] Olof J. Staffans. Quadratic optimal control of well-posed linear systems. To appear in SIAM Journal on Control and Optimization, 1997.

[11] Olof J. Staffans. Feedback representations of critical controls for well-posed linear systems. Submitted, 1997.

[12] Olof J. Staffans. On the distributed stable full information H^∞ minimax problem. Submitted, 1997.

[13] Olof J. Staffans. Quadratic optimal control of regular well-posed linear systems, with applications to parabolic equations. Submitted, 1997.

[14] George Weiss. Transfer functions of regular linear systems. Part I: Characterizations of regularity. *Transactions of American Mathematical Society*, 342:827–854, 1994.

[15] George Weiss. Regular linear systems with feedback. *Mathematics of Control, Signals, and Systems*, 7:23–57, 1994.

[16] George Weiss and Hans Zwart. An example in linear quadratic optimal control. To appear in Systems and Control Letters, 1997.

[17] Martin Weiss and George Weiss. Optimal control of stable weakly regular linear systems. *Mathematics of Control, Signals, and Systems*, 10:287–330, 1997.

20

Convergence of an algorithm for the Riemannian SVD

Bart De Moor

ESAT-SISTA
Katholieke Universiteit Leuven
3001 Leuven
BELGIUM
demoor@esat.kuleuven.ac.be

1 Problem statement

The Riemannian SVD of a given matrix $A \in \mathbf{R}^{p \times q}$ is a nonlinear generalization of the SVD:

Riemannian SVD

$$\begin{aligned} A\,v &= D_v\,u\,\tau\,, & u^T D_v u = 1\,,\;\; v^T v = 1\,, \\ A^T\,u &= D_u\,v\,\tau\,, & v^T D_u v = 1\,. \end{aligned} \tag{20.1}$$

Here $u \in \mathbf{R}^p$ and $v \in \mathbf{R}^q$ are a left, resp. right singular vector and $\tau \in \mathbf{R}$ is a singular value. The matrices $D_u \in \mathbf{R}^{q \times q}$ and $D_v \in \mathbf{R}^{p \times p}$ are symmetric positive definite matrix functions, the elements of which are quadratic in the components of u, resp. v.

The singular triplet (u, τ, v) corresponding to the smallest singular value τ, provides the solution to a so-called stuctured and/or weighted total least squares problems (STLS and/or WTLS), which, for a given data matrix $A \in \mathbf{R}^{p \times q}$ and given nonnegative weights w_{ij} (including the limiting cases 0 and $+\infty$), is the following

$$\min_{B \in \mathbf{R}^{p \times q}\,,\ \mathrm{rank}(B) = q-1\,,\ B \text{ structured}} \sum_{i=1}^{p} \sum_{j=1}^{q} (a_{ij} - b_{ij})^2 w_{ij}\;. \tag{20.2}$$

By 'B structured', we mean so-called affine structures, i.e. $B = B_0 + B_1\beta_1 + \ldots + B_N\beta_N$, where the matrices $B_i \in \mathbf{R}^{p \times q}$ are fixed and the scalars $\beta_i \in \mathbf{R}$ are the parameters. Examples of such structured matrices are (centro- and per-) symmetric matrices, (block-)Hankel, (block-)Toeplitz, circulant, Brownian, Hankel + Toeplitz matrices, matrices with a certain zero structure (sparsity pattern), etc ...[1] Structured/weighted total least squares problems have a wide variety of applications in signal processing and system identification (see e.g. [3] [4] [6] [5] or consult the search engine at http://www.esat.kuleuven.ac.be/sista/publications). The precise structure of the matrices D_u and D_v in the Riemannian SVD depends on the structure of A and B and the given weights. As an example, when B is required to be a $p \times q$ rank deficient Hankel matrix, and when all weights in (20.2) are 1, we have

$$D_u = T_u T_u^T \ , \ D_v = T_v T_v^T \ ,$$

where $T_u \in \mathbf{R}^{p \times (p+q-1)}$ is a banded Toeplitz matrix with the components of u as

$$T_u = \begin{pmatrix} u_1 & u_2 & \cdots & \cdots & u_{p-1} & u_p & 0 & \cdots & 0 & 0 \\ 0 & u_1 & u_2 & \cdots & u_{p-2} & u_{p-1} & u_p & \cdots & 0 & 0 \\ \cdots & \cdots & \cdots & \cdots & \cdots & \cdots & \cdots & \cdots & \cdots & \cdots \\ 0 & 0 & 0 & \cdots & 0 & 0 & u_1 & \cdots & \cdots & u_p \end{pmatrix} \tag{20.3}$$

and $T_v \in \mathbf{R}^{p \times (p+q-1)}$ is constructed similarly from the components of v. We refer to [4] for a derivation and details.
The quest for the minimal singular value τ in the Riemannian SVD, originates in the fact that τ^2 is equal to the objective function in (20.2) (see e.g. [3] [4] for a derivation). For invertible D_v, we have $D_v^{-1} A v = u\tau$, hence the generalized (nonlinear) eigenvalue problem

$$A^T D_v^{-1} A v = D_u v \tau^2 \ . \tag{20.4}$$

Premultiplication with v^T, shows that finding the minimal singular value τ, is equivalent with finding the minimizing solution v of the constrained optimization problem[2]:

$$\min_v \ \tau^2 = v^T A^T D_v^{-1} A v \ \text{ subject to } \ v^T v = 1 \ . \tag{20.5}$$

[1]When B is unstructured, and all weights are 1, we have the well-known *total linear least squares problem* (TLLS), the solution of which is given by the singular value decomposition (SVD) of A (put $D_u = I_q$ and $D_v = I_p$ in (20.1)); For references on TLLS see e.g. [6] [7] [8].

[2]When v is a minimizer, we can find u as follows: Call $z = D_v^{-1} A v$. Then it follows from the objective function (20.5) that $z^T D_v z = \tau^2$. Call $u = z/\tau$. It follows that $Av = D_v u \tau$. Furthermore, we always have the equality $v^T D_u v = u^T D_v u$. Hence the equivalence.

2 A convergent algorithm ?

Of course, since finding the minimal singular value τ of the Riemannian SVD is equivalent with finding the minimum of the constrained nonlinear optimization problem (20.5), one could apply 'classical' optimization techniques such as Gauss-Newton and its variants (see e.g. [1] where STLS / WTLS problems are treated this way). But when applying this approach, we ignore the 'SVD-like' character of the solution. If for instance we calculate the minimal singular value of a matrix, we use a 'dedicated' algorithm for doing so, even if we can formulate that problem as a nonlinear constrained optimization problem. There are indications that suggest that a 'dedicated' algorithm for the Riemannian SVD might be conceivable. Indeed, observe that, for fixed D_u and D_v (i.e. D_u and D_v are not a function of u, resp. v), the Riemannian SVD reduces to the so-called *restricted SVD* described in [2]. For this case, the minimal singular value may be found by the classical century-old (inverse) power method (see e.g. [7]). This simple observation leads to the following proposal for a heuristic power method approach to find the minimal singular value of the Riemannian SVD, which is based on the generalized eigenvalue problem (20.4), in which we fix the matrices D_u and D_v in each iteration step:

(Inverse) power method for Riemannian SVD Initialize:
$u^{[0]}$, $v^{[0]}$, $D_{u^{[0]}}, D_{v^{[0]}}, k = 0.$
Iterate until convergence:
$v^{[k+1]} = (A^T D_{v^{[k]}} A)^{-1} D_{u^{[k]}} v_k$
$v^{[k+1]} = v^{[k+1]} / \|v^{[k+1]}\|$
Update $D_{v^{[k+1]}}$
$u^{[k+1]} = D_{v^{[k+1]}}^{-1} A v^{[k+1]}$
$u^{[k+1]} = u^{[k+1]} / \|(u^{[k+1]})^T D_{v^{[k+1]}} u^{[k+1]}\|$
Update $D_{u^{[k+1]}}$

On numerical try outs, this algorithm seems to work well (we omit implementation details here). It convergence rate seems to be linear. The following questions now apply:

- Can something rigoreus be proven about the convergence behavior and rate of this 'heuristic' method ?

- In particular, the method we propose is like a power method, in which in each iteration the 'metric', as represented by D_u and D_v, is updated. The adaptation of the positive definite matrices D_u and D_v in each iteration step is reminiscent of so-called 'variable metric' methods. Can these observations be exploited in a convergence analysis?

- What about local and global minima ? (Observe that for fixed positive definite D_u and D_v, there is only one minimum, which is global. In this case, the other stationary points of (20.4) and (20.5) are saddle-points).

3 References

[1] Abatzoglou T.J., Mendel J.M., Harada G.A., *The constrained total least squares technique and its application to harmonic superresolution*, IEEE Transactions on Signal Processing, SP-39, no.5, pp. 1070-1087, 1991.

[2] De Moor B., Golub G.H. *The restricted singular value decomposition: properties and applications.* Siam Journal on Matrix Analysis and Applications, Vol.12, no.3, July 1991, pp.401-425.

[3] De Moor B. *Structured total least squares and L_2 approximation problems.* Special issue of Linear Algebra and its Applications, on Numerical Linear Algebra Methods in Control, Signals and Systems (eds: Van Dooren, Ammar, Nichols, Mehrmann), Volume 188-189, July 1993, pp.163-207.

[4] De Moor B. *Total least squares for affinely structured matrices and the noisy realization problem.* IEEE Transactions on Signal Processing, Vol.42, no.11, November 1994.

[5] De Moor B. *Linear system identification, structured total least squares and the Riemannian SVD.* in *Recent Advances in Total Least Squares Techniques and Errors-in-Variables modelling*, ed. S. Van Huffel, SIAM, 1997, pp.225-239.

[6] De Moor B. *Structured total least squares for Hankel matrices.* in 'Communications, Computation, Control and Signal Processing', eds: Arogyaswami Paulraj, Vwani Roychowdhury, Charles Schaper, Kluwer Academic Publishers, 1997, pp.243-258, (at the occasion of Thomas Kailath's 60th birthday, Stanford, USA).

[7] Golub G.H., Van Loan C. *Matrix Computations.* Johns Hopkins University Press, Baltimore 1989 (2nd edition).

[8] Van Huffel S., Vandewalle J. *The Total Least Squares Problem: Computational Aspects and Analysis.* Frontiers in Applied Mathematics 9, SIAM, Philadelphia, 300 pp., 1991.

21

Some open questions related to flat nonlinear systems

Michel Fliess*, Jean Lévine**, Philippe Martin** and Pierre Rouchon***[1]

*Laboratoire des Signaux et Systèmes
CNRS–Supélec–Université Paris-Sud
Plateau de Moulon
91192 Gif-sur-Yvette
FRANCE
fliess@lss.supelec.fr

**Centre Automatique et Systèmes
École des Mines de Paris
35, rue Saint-Honoré
77305 Fontainebleau
FRANCE
levine@cas.ensmp.fr
martin@cas.ensmp.fr

***Centre Automatique et Systèmes
École des Mines de Paris
60, bd. Saint-Michel
75272 Paris Cedex 06
FRANCE
rouchon@cas.ensmp.fr

1 Introduction

(Differentially) flat nonlinear systems were introduced by the authors by employing either differential algebra or the differential geometry of infinite order jets and prolongations (see [4, 7], and the references therein; no-

[1]Work partially supported by the European Commission's Training and Mobility of Researchers (TMR), Contract ERBFMRXT-CT970137, and by the PRC-GDR *Automatique*.

tice, also, the contributions in [13] and [12]). In accordance with Willems' standpoint [16], flatness is best understood without making any distinction between the system variables:

1. Every system variable may be expressed as a function of the components of a finite set $y = (y_1, \ldots, y_m)$ and of a finite number of their time-derivatives.

2. Every component of y may be expressed as a function of the system variables and of a finite number of their time-derivatives;

3. The components of y are not related by any differential relation. When an input of m independent input channels is given, the last condition may be replaced by the following one: the number of components of y is equal to m.

The fictitious output y is called a *flat*, or *linearizing, output.* It may almost always be chosen with a clear engineering and physical meaning. It can be shown that a flat system is equivalent to a controllable constant linear system via a special type of dynamic feedback, called *endogenous.*
The flat output leads to a straightforward solution to the motion planning problem. The equivalence to a linear controllable system although provides a feedback law stabilizing around the reference trajectory. This new setting has already been illustrated by numerous realistic case-studies such as the motion planning and the stabilization of several nonholonomic mechanical systems [4, 5, 7], of a towed cable [11], of an aircraft [9, 7], of magnetic bearings [8], of chemical reactors [14, 15], etc.

2 Flatness characterization

We restrict for simplicity to the classic dynamics

$$\dot{x} = f(x, u) \tag{21.1}$$

where $x \in I\!R^n$, $u \in I\!R^m$. The existence of a general test for checking whether (21.1) is flat is an open problem. The main difficulty is that a candidate flat output $y = h(x, u, \ldots, u^{(r)})$ may *a priori* depend on derivatives of u of arbitrary order r. Whether this order r admits an upper bound (in terms of n and m) is completely unknown. Hence we do not know whether a finite bound exists at all. In the sequel, we say (21.1) is r-flat if it admits a flat output depending on derivatives of u of order at most r.
To illustrate this upper bound might be at least linear in the state dimension, consider the system

$$x_1^{(\alpha_1)} = u_1, \quad x_2^{(\alpha_2)} = u_2, \quad \dot{x}_3 = u_1 u_2$$

with $\alpha_1 > 0$ and $\alpha_2 > 0$. It admits the flat output

$$y_1 = x_3 + \sum_{i=1}^{\alpha_1}(-1)^i x_1^{(\alpha_1-i)} u_2^{(i-1)}, \quad y_2 = x_2,$$

hence is r-flat with $r := \min(\alpha_1, \alpha_2) - 1$. We suspect (without proof) there is no flat output depending on derivatives of u of order less than $r-1$.
If such a bound $\kappa(n,m)$ were known, the problem would amount to checking p-flatness for a *given* $p \leq \kappa(n,m)$ and could be solved in theory. Indeed, it consists [9] in finding m functions $h_1, \ldots, h_m$ depending on $(x, u, \ldots, u^{(p)})$ such that

$$\dim \operatorname{span}\left\{dx_1, \ldots, dx_n, du_1, \ldots, du_m, dh_1^{(\mu)}, \ldots, dh_m^{(\mu)}\right\}_{0 \leq \mu \leq \nu} = m(\nu+1),$$

where $\nu := n + pm$. This means checking the integrability of the partial differential system with a transversality condition

$$\begin{aligned} dx_i \wedge dh \wedge \ldots \wedge dh^{(\nu)} &= 0, \quad i = 1, \ldots, n \\ du_j \wedge dh \wedge \ldots \wedge dh^{(\nu)} &= 0, \quad j = 1, \ldots, m \\ dh \wedge \ldots \wedge dh^{(\nu)} &\neq 0, \end{aligned}$$

where $dh^{(\mu)}$ stands for $dh_1^{(\mu)} \wedge \ldots \wedge dh_m^{(\mu)}$. Its (formal) integrability may be verified by determining *Spencer's cohomology* [1]. Such calculations are in general quite tedious. It can be hoped that, due to the special structure of the above equations, major simplifications might appear.
REMARK. We refer to [10] for more details and a short overview on known results for checking flatness.

3 Willems' controllability

It is quite clear that any differentially flat system locally satisfies Willems' trajectory characterization of controllability [16]. This property is also verified by *orbitally flat* systems [6, 7], where time-scaling is allowed.
How to give a precise definition of Willems' controllability in a general nonlinear setting? What are the nonlinear systems which satisfy this property and in which way are they related to flatness? (See [2, 3] and [4] for an intrinsic algebraic approach of linear controllability which is strongly related to Willems' approach.)

4 References

[1] R.L. Bryant, S.S. Chern, R.B. Gardner, H.L. Goldschmidt and P.A. Griffiths, *Exterior Differential Systems*, Springer, New York, 1991.

[2] M. Fliess, Some basic structural properties of generalized linear systems, *Systems Control Lett.*, **15**, 1991, 391-396.

[3] M. Fliess, A remark on Willems' trajectory characterization of linear controllability, *Systems Control Lett.*, **19**, 1992, 43-45.

[4] M. Fliess, J. Lévine, P. Martin and P. Rouchon, Flatness and defect of nonlinear systems: introductory theory and applications, *Internat. J. Control*, **61**, 1995, 1327-1361.

[5] M. Fliess, J. Lévine, P. Martin and P. Rouchon, Design of trajectory stabilizing feedback for driftless flat systems, *Proc.* 3rd *European Control Conf.*, Rome, 1995, pp. 1882-1887.

[6] M. Fliess, J. Lévine, P. Martin and P. Rouchon, Deux applications de la géométrie locale des diffiétés, *Ann. Inst. H. Poincaré Phys. Théor.*, **66**, 1997, 275-292.

[7] M. Fliess, J. Lévine, P. Martin and P. Rouchon, A Lie-Bäcklund approach to equivalence and flatness of nonlinear systems, *IEEE Trans. Automat. Control*, to appear.

[8] J. Lévine, J. Lottin and J.C. Ponsart, A nonlinear approach to the control of magnetic bearings, *IEEE Trans. Control Systems Technology*, **4**, 1996, 524-544.

[9] P. Martin, *Contribution à l'étude des systèmes différentiellement plats*, Thèse, École des Mines, Paris, 1992.

[10] P. Martin, R.M. Murray, and P. Rouchon, Flat systems, *Plenary Lectures and Mini-Courses ECC 97*, G. Bastin and M. Gevers Eds, Brussels, 1997, pp. 211-264.

[11] R.M. Murray, Trajectory generation for a towed cable flight control system, *Proc. IFAC World Congress*, San Francisco, 1996.

[12] M. van Nieuwstadt, M. Ratinam and R.M. Murray, Differential flatness and absolute equivalence, *Proc.* 33rd *IEEE Conf. Decision Control*, Lake Buena Vista, FL, 1994, pp. 326-332.

[13] J.-B. Pomet, A differential geometric setting for dynamic equivalence and dynamic linearization, in *Geometry in Nonlinear Control and Differential Inclusions*, B. Jakubczyk, W. Respondek and T. Rzeżuchowski Eds, pp. 319-339 (Banach Center Publications, Warsaw, 1995).

[14] R. Rothfuß, *Anwendung der flachbasierten Analyse und Regelung nichtlinearer Mehrgrößoensysteme*, VDI, Düsseldorf, 1997.

[15] R. Rothfuß, J. Rudolph and M. Zeitz, Flatness based control of a nonlinear chemical reactor model, *Automatica*, **32**, 1996, 1433-1439.

[16] J.C. Willems, Paradigms and puzzles in the theory of dynamical systems, *IEEE Trans. Automat. Control*, **36** (1992) 259-294.

22

Approximation of complex μ

Minyue Fu

Department of Electrical and Computer Engineering
University of Newcastle
Newcastle, NSW 2308
AUSTRALIA
eemf@hartley.newcastle.edu.au

1 Description of the problem

Given a matrix $M \in \mathbf{C}^{n\times n}$ and a set of positive integers $\mathcal{X} = (k_1, \cdots, k_m)$ with $k_1 + \cdots + k_m = n$, the so-called *complex structured singular value* $\mu_{\mathcal{X}}(M)$, (complex μ for short), is defined as follows:

$$\mu_{\mathbf{\Delta}}(M) = \inf\left\{\rho : \rho > 0, \det(I_n - \rho^{-1}\Delta M) \neq 0,\ \forall\ \Delta \in B(\mathbf{\Delta})\right\} \quad (22.1)$$

where

$$B(\mathbf{\Delta}) = \left\{\Delta = \mathrm{diag}\{\Delta_1, \cdots, \Delta_m\} \mid \Delta_i \in \mathbf{C}^{k_i \times k_i}, ||\Delta_i|| \leq 1\right\} \quad (22.2)$$

Let $\hat{\mu}$ be an approximation of μ. We call $\hat{\mu}$ an r-approximation, $r > 0$, if either

$$\mu \leq \hat{\mu} \leq (1+r)\mu \quad (22.3)$$

or

$$\frac{\mu}{1+r} \leq \hat{\mu} \leq \mu \quad (22.4)$$

Note that $\hat{\mu}$ is an upper bound in the former case and an lower bound in the latter.
We are interested in the computational complexity of the problem of approximating the complex μ. More specifically, we ask the following questions:

1. Does there exist some (arbitrarily small) constant $\epsilon > 0$ such that the problem of finding an ϵ-approximation for the complex μ is NP-hard?

2. For any (arbitrarily large) constant $R > 0$, is the problem of finding an R-approximation for the complex μ is NP-hard?

2 Motivations

The complex μ problem arises in robustness stability and robust performance problems where systems uncertainties and performance measures can be captured by the structure of $\mathcal{D}$. This problem was first formally proposed by Doyle in 1982 [1] where the so-called D-scaling method was introduced for computing an upper bound for μ. Many simulation results have been conducted by Doyle and his colleagues to demonstrate that the D-scaling method provides good approximaitons (with relative error no larger than 20%). However, there is no theoretical result supporting the simulation results. It is worth to know that the D-scaling method gives a polynomial algorithm.

In 1993, Megretski [2] showed that the D-scaling method gives a relative error r which grows at most linearly as the function of n, provided that $\mu \neq 0$. That is, the problem of finding a linearly growing r-approximation for the complex μ is a polynomial problem, provided $\mu \neq 0$.

The computational complexity of the μ problem became of interest since early 90's. A number of authors studied the so-called *real* μ problem where the complex blocks D_i are replaced with the so-called *repeated real blocks* $\delta_i I_{k_i}$, $\delta_i \in \mathcal{R}$, and the so-called *mixed* μ problem where there are real repeated blocks, complex blocks, and the so-called *repeated complex blocks* which are similar to the repeated real blocks except δ_i are complex variables. Coxson and DeMarco [3] (among many other people) showed that the real μ problem (and hence the mixed μ problem) is NP-hard. More specifically, the problem of determining if $\mu < 1$ is NP-complete. Subsequently, there exists no polynomial algorithm for computing μ, unless the commonly believed conjecture, $P \neq NP$, fails. The computational complexity of the complex μ is much harder to analyze than the real μ problem. Recently, Toker and Ozbay [4] used an elegant technique to show that the complex μ problem is still NP-hard.

Knowing that the problem of computing μ is NP-hard, the next logical question is how "hard" it is to approximate μ. To this end, a result in Coxson and DeMarco [3] shows that there exists some arbitrarily small $\epsilon > 0$ such that ϵ-approximation for the real μ is also NP-hard. Toker [5] offers a more negative answer for the real μ problem by proving that computing a $Cn^{1-\epsilon}$-approximation with some (very large) constant $C > 0$ and (very small) $\epsilon > 0$ is an NP-hard problem. Fu in a very recent paper [6] gives the following most negative result: The problem of $r(n)$-approximation for the real μ is NP-hard for any $r(n) > 0$. That is, the realtive approximation error can grow arbitrarily fast for any polynomial algorithm, unelss $P = NP$. The result in [6] is further extended by Fu and Dasgupta [7] to the case where the real blocks are bounded using an arbitrary p-norm rather than ∞-norm and imilar NP-hard results are obtained.

However, the computational complexity analysis for approximation of the complex μ appears to be much more involved. Hence, we invite the reader to study the proposed questions. We conjecture that the answers to both questions are affirmative. The first question appears to be easier to answer than the second one. So we rank the first question "moderate" and the second one "difficult".

3 References

[1] J. C. Doyle, "Analysis of Feedback Systems with Structured Uncertainties," *Proc. IEE*, pt. D, vol. 129, pp. 240-250, 1982.

[2] Megretski, A., "On the gap between structured singular values and their upper bounds," *Proc. 32nd Conf. Decision and Control*, San Antonio, TX, December, 1993.

[3] G. E. Coxson and C. L. DeMarco, "The computational complexity of approximating the minimal perturbation scaling to achieve instability in an interval matrix," *Mathematics of Control, Signals, and Systems*, **7** (No. 4), pp. 279-292, 1994.

[4] O. Toker and H. Ozbay, "On the Complexity of Purely Complex μ Computation and Related Problems in Multidimensional Systems," *IEEE Trans. Auto. Contr.*, vol. 43, no. 3, pp. 409-414, 1998.

[5] O. Toker, "On the conservatism of upper bound tests for structured singular value analysis," *Proc. 35th Conference on Decision and Control*, Kobe, Japan, Dec. 1996, pp. 1295–1300.

[6] M. Fu, "The real structured singular value is hardly approximable," *IEEE Trans. Auto. Contr.*, vol. 42, no. 9, pp. 1286-1288, 1997.

[7] M. Fu and S. Dasgupta, "Computational complexity of real structured singular value in *p*-norm Setting," to appear in *IEEE Trans. Auto. Contr.*

23

Spectral value sets of infinite-dimensional systems

E. Gallestey*, D. Hinrichsen* and A. J. Pritchard**

*Institut für Dynamische Systeme
Universität Bremen
D-28334 Bremen
GERMANY
dh@mathematik.uni-bremen.de

**Mathematics Department
University of Warwick
Coventry CV4 7AL
UNITED KINGDOM
ajp@maths.warwick.ac.uk

1 Description of the problem

We assume that $\underline{X}$, X, U, Y are complex or real separable Banach spaces, A with dense domain $D(A) \subset X$ is a closed linear operator on X, $D(A) \subset \underline{X} \subset X$ with continuous dense injections, $B \in \mathcal{L}(U, X)$ and $C \in \mathcal{L}(\underline{X}, Y)$. Let $\mathbf{K}$ denote the field of scalars. Our subject is the variation of the spectrum, $\sigma(A)$ under structured perturbations of the form

$$A \rightsquigarrow A_\Delta = A + B\Delta C, \quad \Delta \in \mathcal{L}(Y, U) \tag{23.1}$$

where $D(A_\Delta) = D(A)$. The operators B, C are fixed and describe both the structure and unboundedness of the perturbations, whilst $\Delta \in \mathcal{L}(Y, U)$ is arbitrary. If $U = X = \underline{X} = Y$, $B = I_X = C$, i.e. $A_\Delta = A + \Delta$, $\Delta \in \mathcal{L}(X)$ the perturbations are bounded and are said to be *unstructured*.

Definition. Given A, B, C as above, the associated *spectral value set* of level $\delta > 0$ is

$$\sigma(A, B, C; \delta) = \{\lambda \in \mathbb{C};\ \exists\, \Delta \in \mathcal{L}(Y, U) :$$
$$\|\Delta\| < \delta\ A_\Delta \text{ is closed and } \lambda \in \sigma(A_\Delta)\}. \tag{23.2}$$

The corresponding *point spectral value set* of level δ, is

$$\sigma_P(A,B,C;\delta) = \{\lambda \in \mathbb{C};\ \exists \Delta \in \mathcal{L}(Y,U) : \|\Delta\| < \delta, A_\Delta \text{ is closed and } \lambda \in \sigma_P(A_\Delta)\}, \tag{23.3}$$

where $\sigma_P(A_\Delta)$ denotes the point spectrum of A_Δ.

In order to eliminate the closedness condition for A_Δ in this definition, a solution of the following problem would be helpful.

Problem 1: Closedness Radius. Determine the closedness radius

$$R_{\text{cl}}(A,\ B,\ C) = \inf\{\|\Delta\|;\ \Delta \in \mathcal{L}(Y,U) : A_\Delta \text{with domain } D(A) \text{ is not closed}\}.$$

In most applications an analytical determination of $\sigma(A)$ and $\sigma(A,B,C;\delta)$ will not be possible. Then it is a natural idea to approximate these sets (within a compact subset of interest in $\mathbb{C}$) via finite dimensional approximations of the operators [2]. More precisely, assume that we have sequences of finite dimensional linear subspaces U_N, X_N, Y_N in $U, D(A), Y$, respectively, such that

$$\begin{aligned}
&\lim_{N\to\infty} \operatorname{dist}(x, X_N) = 0\\
&\lim_{N\to\infty} \operatorname{dist}(u, U_N) = 0\\
&\lim_{N\to\infty} \operatorname{dist}(y, Y_N) = 0\\
&x \in X,\ u \in U,\ y \in Y.
\end{aligned}$$

We assume that $\dim X = \infty$, $\dim X_N = N$, but we do not exclude the possibility that U and/or Y are finite dimensional. In this case the sequence of subspaces (U_N) and/or (Y_N) will become stationary. Usually the subspaces U_N, X_N, Y_N will be defined via some bases. E.g. one sets

$$X_N = \operatorname{span}\{a_1^{(N)}, \ldots, a_N^{(N)}\}, \quad N \in \mathbb{N}$$

where $a_1^{(N)}, \ldots, a_N^{(N)}$ is an independent family of N vectors in $D(A)$. We define the triplets (A_N, B_N, C_N) of finite dimensional operators by

$$A_N = \pi_N A\, \iota_N, \quad B_N = \pi_N B\, \iota_N^U, \quad C_N = \pi_N^Y C\, \iota_N, \qquad N \in \mathbb{N}$$

where $\iota_N : X_N \to D(A)$ is the canonical injection and $\pi_N : X \to X_N$ a suitable projection. Similarly for U and Y.

Problem 2: Approximation of Spectral Value Sets. Under which conditions do we have convergence

$$\sigma(A_N, B_N, C_N;\delta) \cap K \to \sigma(A,B,C;\delta) \cap K \quad \text{as } N \to \infty$$

(with respect to the Hausdorff metric), for every compact subset K of the resolvent set $\rho(A)$?

In view of the available results for the complex case (see Section 3), this leads to the following

Problem 3: Approximation of transfer functions. Under which conditions does $\|C_N(sI_N - A_N)^{-1}B_N\|$ converge uniformly on compact subsets of $\rho(A)$ towards $\|C(sI - A)^{-1}B\|$? (This implies in particular, that for all compact $K \subset \rho(A)$ one has $K \subset \rho(A_N)$ for N sufficiently large).

2 Motivations

Spectral analysis plays a basic role in many fields of applied mathematics. In recent years the idea of spectral value sets has been developed independently in Numerical Analysis and in Control Theory. The concept has proved to be of interest in the following areas:

1. *Analysis and control of uncertain linear systems.* In control systems design an indispensible requirement is stabilization, but in addition other performance requirements are often specified in terms of the location of poles of the feedback system. In the presence of uncertainty one should make sure that not only the spectrum of the nominal feedback system but also the spectra of all systems lying in a neighbourhood reflecting the model uncertainty lie in the prescribed region of the complex plane. For finite dimensional systems spectral value sets for *unstructured real* perturbations have been analyzed in [9] and for *structured complex* perturbations in [7].

2. *Numerical Analysis.* It is well known that if an operator is not normal its spectrum may be highly sensitive to small parameter perturbations. Then the meaning of the spectrum becomes doubtful in the presence of rounding errors and small model inaccuracies. Motivated by this problem of numerical spectral analysis, the idea of spectral value sets (for *unstructured complex* perturbations) has been pursued by Trefethen [14], [15] under the name of *pseudospectra* and by Godunov [5] under the name of *spectral portraits.*

3. *Stability analysis in Fluid Dynamics.* Reddy and his co-workers have computed the spectral value sets of the Orr-Sommerfeld operator [12] (for unstructured complex perturbations). The sets show that the spectrum is very sensitive to perturbations and this has been put forward as a reason for experimental results which show that a Poiseuille flow becomes unstable at values of the Reynolds number which are much smaller than those theoretically predicted by a spectral analysis of the Orr-Sommerfeld operator [12]. However starting

from the Navier-Stokes equations and following the usual route to the Orr-Sommerfeld equation, an adequate representation of the effect of nonlinearities and parametric uncertainties requires the analysis of bounded or unbounded *structured* perturbations, see [6].

3 Available results

1. Kato [11] has shown that perturbations which are A-bounded with A-bound less than one preserve the property of closedness.

2. In the real case where the Banach spaces $\underline{X}$, X, U, Y are all real ($\mathbf{K} = \mathbb{R}$) the analysis of spectral value sets is more complicated and, to our knowledge, there are only finite dimensional results available, see [9], [8].

3. If A, B, C are matrices over the ground field $\mathbf{K} = \mathbb{C}$, the spectral value set can be characterized via the superlevel sets of $\|G\|$ where $G(s) = C(sI - A)^{-1}B$ is the transfer function associated with (A, B, C) [7]:
$$\sigma(A, B, C; \delta) = \sigma(A) \cup \{s \in \mathbb{C} \setminus \sigma(A);\ \|G(s)\| > \delta^{-1}\}.$$
Under certain conditions this characterization can be extended to the infinite dimensional context, see [4]. However, in general, the transfer function of an infinite dimensional system cannot be obtained analytically, so the extended characterization does not usually enable us to determine $\sigma(A, B, C; \delta)$ directly.

4. There are efficient algorithms to compute spectral value sets in the complex finite dimensional case, see [3] and the references therein.

5. For Toeplitz operators $T = T(a) : l^2(\mathbb{N}, \mathbb{C}) \to l^2(\mathbb{N}, \mathbb{C})$ with piecewise continuous symbol $a : \mathbf{T} \to \mathbb{C}$ ($\mathbf{T} = \{s \in \mathbb{C};\ |s| = 1\}$) and associated infinite matrix $T = (t_{i-j})_{i,j \in \mathbb{N}}$ it has been shown that the *spectra* of the finite dimensional truncations $T_N = (t_{i-j})_{i,j \in \underline{N}}$ ($\underline{N} = \{1, \ldots, N\}$) do *not*, in general, approximate the spectrum of T whereas, for any $\delta > 0$, the *unstructured spectral value sets* $\sigma(T_N, I_N, I_N; \delta)$ of T_N converge towards the corresponding unstructured spectral value set $\sigma(T, I, I; \delta)$ of T, see [1].

6. In the case of delay systems Problem 2 can be resolved using spline-based schemes, see [13], [10].

4 References

[1] A. Böttcher. Pseudospectra and singular values of large convolution operators. *J. Integral Equations and Applications* 6: 267–301, 1994.

[2] F. Chatelin. Spectral Approximation of Linear Operators. Academic Press, London, 1983.

[3] E. Gallestey. Computing spectral value sets using the subharmonicity of the norm of rational matrices. *BIT* 38: 22–33, 1998.

[4] E. Gallestey, D. Hinrichsen, and A. J. Pritchard. On the spectral value set problem in infinite dimensions. In *Proc. 4rd International Symposium on Methods and Models in Automation and Robotics*, Miedzyzdroje (Poland), pp 109–114, 1997.

[5] S. K. Godunov. Spectral portraits of matrices and criteria of spectrum dichotomy. In *Proc. of Internat. Conf. on Computer Arithmetic, Scientific Computation and Mathematical Modelling. SCAN-91*, Oldenburg, 1991.

[6] T. Herbert. On perturbation methods in nonlinear stability theory. *J. Fluid Mech.*, 126:167–186, 1983.

[7] D. Hinrichsen and B. Kelb. Spectral value sets: a graphical tool for robustness analysis. *Systems & Control Letters*, 21: 127–136, 1993.

[8] D. Hinrichsen and B. Kelb. Stability radii and spectral value sets for real matrix perturbations. In *Systems and Networks: Mathematical Theory and Applications, Vol.II Invited and Contributed Papers*, Akademie-Verlag, Berlin, 217–220, 1994.

[9] D. Hinrichsen and A. J. Pritchard. On spectral variations under bounded real matrix perturbations. *Numerische Mathematik* 60: 509–524, 1992.

[10] K. Ito and F. Kappel A uniformly differentiable approximation scheme for delay systems using splines. Applied Mathematics and Optimization 23: 217-262, 1991.

[11] T. Kato. *Perturbation Theory for Linear Operators.* Springer-Verlag, Berlin-Heidelberg-New York, 1976.

[12] S. C. Reddy, P. Schmidt, and D. Henningson. Pseudospectra of the Orr-Sommerfeld operator. *SIAM J. Appl. Math.*, 53: 15–47, 1993.

[13] D. Salamon. Structure and stability of finite dimensional approximations for functional differential equations. *SIAM J. Contr. & Opt.* 23: 928–951, 1985.

[14] L. N. Trefethen. Pseudospectra of matrices. Report 91/10, Oxford University Computer Laboratory, 1991.

[15] L. N. Trefethen. Pseudospectra of linear operators. SIAM Rev. 39: 383-406, 1997.

24
Selection of the number of inputs and states

Christiaan Heij

Econometric Institute
Erasmus University Rotterdam
P.O. Box 1738
3000 DR Rotterdam
THE NETHERLANDS
heij@few.eur.nl

1 Problem statement

Suppose that a collection of q real-valued variables, denoted by w, is observed in discrete time Z. The basic problem in system identification is to construct a system $\mathcal{B}$ that reflects the dynamic relations between these variables. Restricting the attention to linear systems, one of the principal choices are the number m of inputs and the number n of states of the system. For example, consider a stochastic input-output system $y = Gu + H\varepsilon$, where ε is a white noise process and G and H are rational transfer functions of sizes $p \times m$ and $p \times p$ respectively, with $p + m = q$. The system variables are $w = (u, y)$, and m is equal to the number of unrestricted variables (the inputs u) and n is equal to the sum of the McMillan degrees of G and H. In terms of a (minimal) state space model, the system can be described by n state variables x_t as

$$x_{t+1} = Ax_t + Bu_t + E\varepsilon_t,$$

$$y_t = Cx_t + Du_t + F\varepsilon_t.$$

We call (m, n) the complexity of the system. Now suppose that for given data w and given complexity (m, n) the system is identified by minimizing a criterion function $e(w, \mathcal{B})$ over the set $\mathbf{B}_{m,n}$ of all linear systems $\mathcal{B}$ that have m inputs and n states. Let the minimal value be denoted by $C_{m,n} = \inf \{e(w, \mathcal{B}); \mathcal{B} \in \mathbf{B}_{m,n}\}$. In general, the values $C_{m,n}$ will be smaller for larger m and n, because the system then imposes less restrictions upon the variables. We state the following general problem.

Problem : Determine (m, n) *from the values* $C_{m,n}$.

This problem allows many possible further specifications, depending on

- **A1** the assumptions on the data w;
- **A2** the choice of the criterion function e;
- **A3** the trade-off between complexity (m, n) and fit $C_{m,n}$.

Various solutions have been proposed for different specifications of **A1, A2** and **A3**. This is partially reviewed in the following section. However, the joint estimation of the complexity (m, n) seems to be an unsolved question. In the final section we discuss a specification of particular interest.

2 Significance tests and order selection

A well-known specification of our general problem is the following.

A1 The data are generated by a stationary stochastic process, so that $m = 0$. For simplicity we assume that $q = 1$.
A2 The model class consists of autoregressive systems of order n, and the criterion is the mean squared prediction error, that is,

$$C_n = \min_{\alpha_j} \frac{1}{N} \sum_{t=n+1}^{N} [w(t) - \sum_{j=1}^{n} \alpha_j w(t-j)]^2,$$

where α_j are real-valued and N is the number of observations.
A3 The choice of n is based on the relative changes $\Delta_n = (C_{n-1} - C_n)/C_n$.

Using the conventional approach of statistical testing, under appropriate assumptions on the data and for large enough N the critical value (with 5% significance) for Δ_n is approximately $4/N$. This can be derived from the t-test (or more general, from the likelihood ratio test) on the significance of the coefficient α_n. One can follow a top-down approach (start with large n and decrease till $\Delta_n > 4/N$) or a bottom-up approach (start with $n = 0$ and increase till $\Delta_n < 4/N$). An alternative is to use information criteria, for example $AIC_n = \log(C_n) + 2n/N$ or $BIC_n = \log(C_n) + n\log(N)/N$. For large enough N, this corresponds to using Δ_n with critical values, respectively, $2/N$ and $\log(N)/N$. Statistical tests and information criteria have also been developed for multivariate systems ($q > 1$) and for ARMA models, see [1, 10, 12, 14].

A second well-known specification of our general problem concerns non-dynamic systems.

A1 The data consists of a random sample from a population; there is no dynamics, so that $n = 0$.
A2 The model class consists of linear subspaces of dimension m.
A3 The choice of m is based on statistical tests.

Different criterion functions and test statistics have been developed, depending on the further statistical specification of the data generating process, in particular the error structure of the data. Well-known examples are principal components and factor analysis models, see [2, 3, 6]. For example, in principal components the relation $y_t = Du_t + F\varepsilon_t$ is reformulated in terms of $w_t = (u_t', y_t')'$ as $w_t = w_t^a + w_t^e$, where $w_t^a \in \mathcal{B} = \ker(-D, I)$ lies in a subspace of dimension m and the criterion function is given by $e(w, \mathcal{B}) = \min\{\sum_{t=1}^N \|w_t^e\|^2; w_t = w_t^a + w_t^e, w_t^a \in \mathcal{B}\}$. This has been extended to stationary processes, where m is estimated point-wise in the frequency domain and n is left unrestricted (in general the resulting model has $n = \infty$), see [4, 5, 13, 16].

Summarizing, our general problem has been investigated for special specifications, in particular for stochastic data and quadratic criteria. Several other approaches have been developed. We mention Hankel norm and l_2 approximation of deterministic input-output systems, see [7, 8], and subspace identification techniques, see [15]. These methods consider model errors as the result of approximation, not of randomness. Further, m is assumed given and the choice of n is based on qualitative criteria, not on the optimization of an explicit criterion function in terms of n.

The assumption that m is given is quite standard in the systems literature. This may sometimes be a valid assumption, but in other situations it is less natural, for example in the case of feedback, or when variables are simultaneously dependent as in econometrics. In such situations the question whether a system variable is free or restricted is more a matter of degree than of kind.

3 Research questions

We describe one specification of the general problem in more detail, related to behavioural identification of systems, see [9, 11]. This concerns the simultaneous estimation of m and n of deterministic systems that approximate the data. The model class consists of all (controllable) linear systems, interpreted as sets of trajectories of the system variables, see [17]. That is, the state equations $y_t = Cx_t + Du_t$, $x_{t+1} = Ax_t + Bu_t$ define a set $\mathcal{B}$ of trajectories $w_t = (u_t', y_t')'$, where $w : Z \to R^q$ belongs to $\mathcal{B}$ if and only if there exists a state trajectory $x : Z \to R^n$ such that the state equations are satisfied. The criterion is defined by

$$C_{m,n} = \inf_{\mathcal{B}} \frac{1}{N} \sum_{t=1}^{N} \|w(t) - w^a(t)\|^2$$

where $w^a \in \mathcal{B}$, a linear system with m inputs and n states. Algorithms for the computation of $C_{m,n}$ are given in [9, 11]. There holds $C_{m_1,n} \leq C_{m_2,n}$ for $m_1 > m_2$ and $C_{m,n_1} \leq C_{m,n_2}$ for $n_1 > n_2$. The research questions are the following.

- **Q1** Formulate data assumptions that are relevant and tractable.
- **Q2** For given data assumptions, derive explicit methods for the selection of (m, n) from the criterion values $C_{m,n}$.
- **Q3** Extend this to other criterion functions.

Assuming that the data are generated by a stationary stochastic process, this problem can be seen as an extension of information criteria to the case where also the number of inputs has to be estimated. However, it is also of interest to formulate a mathematical framework that treats the models as approximations without randomness. Answers to the above questions may serve as a basis to compare existing approximation methods in system identification.

4 References

[1] H. Akaike, "A New Look at the Statistical Model Identification", *IEEE Trans. Aut. Control*, 19, pp. 716–723 (1974).

[2] T.W. Anderson, *An Introduction to Multivariate Statistical Analysis*, Wiley, New York, 1984.

[3] T.W. Anderson and H. Rubin, "Statistical Inference in Factor Analysis", in J. Neyman (ed.), *Proc. Third Berkeley Symp. Math. Stat. Prob.*, Vol. 5, Un. California, pp. 111-150 (1956).

[4] D.R. Brillinger, *Time Series: Data Analysis and Theory*, Holt, Rinehart and Winston, New York, 1975.

[5] J.F. Geweke, "The Dynamic Factor Analysis of Economic Time Series Models", in D.J. Aigner and A.S. Goldberger (eds.), *Latent Variables in Socio-economic Models*, North Holland, Amsterdam, pp. 365-383 (1977).

[6] L.J. Gleser, "Estimation in a Multivariate Errors in Variables Regression Model: Large Sample Results", *Ann. Statistics*, 9, pp. 24-44 (1981).

[7] K. Glover, "All Optimal Hankel Norm Approximations of Linear Multivariable Systems and their $\mathcal{L}_\infty$-Error Bounds", *Int. J. Control*, 39, pp. 1115-1193 (1984).

[8] B. Hanzon and J.M. Maciejowski, "Constructive Algebra Methods for the L_2-Problem for Stable Linear Systems", *Automatica*, 32, pp. 1645-1657 (1996).

[9] C. Heij and W. Scherrer, "Consistency of System Identification by Global Total Least Squares", *Report* 9635, Econometric Inst., Erasmus Un. Rotterdam, 1996.

[10] H. Lütkepohl, *Introduction to Multiple Time Series Analysis* (2-nd ed.), Springer, Berlin, 1993.

[11] B. Roorda, *Global Total Least Squares*, Tinbergen Inst. Res. Ser. 88, Thesis Publ., Amsterdam, 1995.

[12] G. Schwarz, "Estimating the Dimension of a Model", *Ann. Statistics*, 6, pp. 461-464 (1978).

[13] D. Stemmer, *Testing the Number of Equations in Linear Dynamic EV-Models*, Ph.D. Thesis, Techn. Un. Vienna, 1995.

[14] P.H.F.M. Van Casteren, *Statistical Model Selection Rules*, Ph.D. Thesis, Free Un. Amsterdam, 1994.

[15] P. Van Overschee and B. De Moor, *Subspace Identification, Theory - Implementation - Application*, Kluwer, Dordrecht, 1995.

[16] C. Weber, *Estimation and Testing of the Number of Factors in Linear Dynamic Factor Analysis Models*, Ph.D. Thesis, Techn. Un. Vienna, 1993.

[17] J.C. Willems, "Paradigms and Puzzles in the Theory of Dynamical Systems", *IEEE Trans. Aut. Control*, 36, pp. 259-294 (1991).

25

Input-output gains of switched linear systems

J. P. Hespanha and A. S. Morse[1]

Center for Computational Vision and Control
Yale University
New Haven, CT 06520
USA
morse@sysc.eng.yale.edu

1 Description of the problem

The problems which follow are motivated primarily by a desire to develop a bona fide performance-based theory of adaptive control. At this point it seems fairly clear that such a theory will require a much better understanding of certain types of linear time-varying systems than we have at present. In an adaptive context, switched {as opposed to continuously tuned} linear systems seem to be the most tractable and it is for this reason that the problems formulated here are for switched linear systems.

Let $\mathcal{P}$ be either a finite or compact subset of a real, finite dimensional linear space and let $M_p : \mathcal{P} \to \mathbb{R}^{n \times n}$, $D_p : \mathcal{P} \to \mathbb{R}^{n \times n_u}$, and $H_p : \mathcal{P} \to \mathbb{R}^{n_y \times n}$ be given continuous functions. Assume that for each $p \in \mathcal{P}$, M_p is a stability matrix. Let τ_D be a given positive number. Let $\mathcal{S}$ denote the set of all piecewise-constant signals $\sigma : [0, \infty) \to \mathcal{P}$ such that either (a) σ switches at most once or (b) σ switches more than once and the time difference between each two successive switching times is bounded below by τ_D.

For each σ in $\mathcal{S}$, the preceding defines a time-varying linear system of the form

$$\Sigma_\sigma \triangleq \left\{ \begin{array}{rcl} \dot{x} & = & M_\sigma x + D_\sigma u \\ y & = & H_\sigma x \end{array} \right\}$$

where u is an integrable input signal taking values in $\mathbb{R}^{n_u}$. Thus if $x(0) \triangleq 0$,

[1]This research was supported by the Air Force Office of Scientific Research, the Army Research Office, and the National Science Foundation.

then $y = Y_\sigma(u)$, where Y_σ is the input-output mapping

$$u \longmapsto \int_0^t H_{\sigma(t)} \Phi_\sigma(t, \tau) D_{\sigma(\tau)} d\tau,$$

and Φ_σ is the state transition matrix of M_σ. Let prime denote transpose and, for any integrable, vector-valued signal v on $[0, \infty)$, let $|| \cdot ||$ denote the two-norm

$$||v|| \triangleq \sqrt{\int_0^\infty v'(t)v(t)dt}$$

Write $\mathcal{L}_2$ for the space of all signals with finite two-norms. The input-output gain of Σ_σ is then the induced two-norm

$$\mu_\sigma \triangleq \inf\{\gamma : ||Y_\sigma(u)|| \leq \gamma ||u||, \; \forall u \in \mathcal{L}_2\}$$

Define the gain μ of the *multi-system* $\{(H_p, M_p, D_p), p \in \mathcal{P}\}$ to be

$$\mu \triangleq \sup_{\sigma \in \mathcal{S}} \mu_\sigma$$

Main Problem: *Derive conditions in terms of* τ_D *and the multi-system* $\{(H_p, M_p, D_p), p \in \mathcal{P}\}$ *under which* μ *is a finite number. Assuming these conditions hold, characterize* μ *in terms of* $\{(H_p, M_p, D_p), p \in \mathcal{P}\}$ *and* τ_D.

The problem just posed implicitly contains as a sub-problem the question of whether or not the time varying matrix M_σ is exponentially stable for every $\sigma \in \mathcal{S}$. This sub-problem, in turn, is arguably the most important open problem in the theory of switched linear systems. Now it is well-known that M_σ will be exponentially stable for each $\sigma \in \mathcal{S}$ if either τ_D is sufficiently large or if the diameter of $\mathcal{P}$ is sufficiently small [1, 2]. Some recent but special results on switched linear systems can be found in [3, 4, 5, 6, 7, 8] and the references therein. In addition, there is an extensive literature on the stability of time varying matrices {see for example pages 116–119 of [9], pages 239–250 of [10], the important survey paper [11] and the "preliminaries" of [12]} but these results are not specific to switched linear systems. Apart from these, very little of a general nature is known about the stability of M_σ, to say nothing of how to characterize μ. It is of course quite easy to calculate stability bounds for τ_D and bounds for μ in terms of normed values of the multi-system's matrices, but these bounds are invariably state-coordinate dependent. They ought not be.

As stated, the main problem is extremely challenging. Moreover while the problem is formulated in the state space, μ and the stability of M_σ are invariant under state coordinate changes of the form $\{(H_p, M_p, D_p), p \in \mathcal{P}\} \longmapsto \{(H_pT^{-1}, TM_pT^{-1}, TD_p), p \in \mathcal{P}\}$ where T is a constant nonsingular matrix. One would thus like to see the main problem reformulated in

a way which takes this invariance into account. We will now formulate a simpler problem which accomplishes this.
Let us consider the case when for $p \in \mathcal{P}$, H_p and D_p are row and column vectors h_p and d respectively, with d not depending on p. Let us restrict M_p to be of the form $M_p \triangleq A + dc_p + bf_p$ where A is a constant matrix, b is a constant n-vector and for $p \in \mathcal{P}$, c_p and f_p are row vectors such that $A + dc_p + bf_p$ is exponentially stable. Thus we are restricting attention to a multi-system of the form $\{(h_p, A + dc_p + bf_p, d), p \in \mathcal{P}\}$. Multi-systems such as this arise quite commonly in adaptive control [2]. For simplicity, suppose that $\mathcal{P}$ is the finite set $\mathcal{P} \triangleq \{1, 2, \ldots, m\}$. Suppose, in addition, without loss of generality, that $(A, [\,b \quad d\,])$ is controllable and that (C, A) is observable, where

$$C = [\,f_1' \quad c_1' \quad h_1' \quad f_2' \quad c_2' \quad h_2' \quad \cdots \quad f_m' \quad c_m' \quad h_m'\,]'$$

Under these conditions the multi-system $\{(h_p, A + dc_p + bf_p, d), p \in \mathcal{P}\}$ uniquely determine a $3m \times 2$ strictly proper rational matrix $W(s) \triangleq C(sI - A)^{-1}\,[\,b \quad d\,]$ of McMillan Degree n. In the light of classical realization theory, it is easy to see that by reversing steps, $W(s)$ uniquely determines $\{(h_p, A + dc_p + bf_p, d), p \in \mathcal{P}\}$ up to a similarity transformation of the form

$$\{(h_p, A + dc_p + bf_p, d),\ p \in \mathcal{P}\} \longmapsto$$

$$\{(h_p T^{-1}, T(A + dc_p + bf_p)T^{-1}, Td),\ p \in \mathcal{P}\}$$

Since neither the stability of $A + dc_\sigma + bf_\sigma$ nor the multi-system gain μ depend on T, we can meaningfully formulate the following problem.

Simplified Problem: *Derive conditions in terms of τ_D and the multi-system rational matrix $W(s)$ under which μ is a finite number. Assuming these conditions hold, characterize μ in terms of $W(s)$ and τ_D.*

Without destroying its significance, the preceding problem can be simplified further by restricting σ to be in the class of "clocked" switching signals $\mathcal{S}_C$ consisting of those switching signals in $\mathcal{S}$ for which the switching times are elements of the infinite sequence $0,\ \tau_D,\ 2\tau_D,\ 3\tau_D, \ldots$[2] One might also try to determine the limiting behavior of μ as $\tau_D \to \infty$. It is for example known that in the limit, μ equals the maximum over $\mathcal{P}$ of the $\mathcal{H}^\infty$ norm of the transfer function $h_p(sI - A - dc_p - bf_p)^{-1}d$ [13].

---

[2]Although further restricting σ to be periodic would make things much more tractable, it would at the same time destroy completely the significance of the problem to adaptive control.

2 References

[1] R. Bellman. *Stability Theory of Differential Equations.* Dover Publications, 1969.

[2] A. S. Morse. Control using logic-based switching. In A. Isidori, editor, *Trends in Control*, pages 69–113. Springer-Verlag, 1995.

[3] K. S. Narendra and J. Balakrishnan. A common Lyapunov function for stable LTI systems with commuting A-matrices. *IEEE Trans. Automatic Control*, 39(12):2469–2471, December 1994.

[4] L. Gurvits. Stability of linear inclusions—part 2. Technical report, NECI, December 1996.

[5] C. F. Martin and W. P. Dayawansa. On the existence of a Lyapunov function for a family of switching systems. In *Proc. of the 35th Conf. on Decision and Contr.*, December 1996.

[6] R. N. Shorten and K. S. Narendra. A sufficient conditions for the existence of a common Lyapunov function for two second order linear systems. In *Proc. of the 36th Conf. on Decision and Contr.*, December 1997.

[7] T. Mori Y. Mori and Y. Kuroe. A solution to the common Lyapunov function problem for continuous-time systems. In *Proc. of the 36th Conf. on Decision and Contr.*, December 1997.

[8] D. Liberzon, J. P. Hespanha, and A. S. Morse. Stability of switched linear systems: a lie-algebraic condition. *System and Control Letters*, 1998. submitted.

[9] W. J. Rugh. *Linear System Theory.* Prentice-Hall, 1993.

[10] H. K. Khalil. *Nonlinear Systems.* Macmillan, 1992.

[11] V. Solo. On the stability of slowly time-varying linear systems. *Systems and control Letters*, pages 331–350, 1994.

[12] F. M. Pait and A. S. Morse. A cyclic switching strategy for parameter-adaptive control. *IEEE Transactions on Automatic Control*, 39(6):1172–1183, June 1994.

[13] J. P. Hespanha. A bound for the induced norm of a switched linear system in terms of a bound on the infinity norms of the transfer functions of the constant linear systems being switched. Technical report, Yale University, March 1998.

26

Robust stability of linear stochastic systems

D. Hinrichsen* and A. J. Pritchard**

*Institut für Dynamische Systeme
Universität Bremen
D-28334 Bremen
GERMANY
dh@mathematik.uni-bremen.de

**Mathematics Department
University of Warwick
Coventry CV4 7AL
UNITED KINGDOM
ajp@maths.warwick.ac.uk

1 Background

Consider a linear system whose parameters are disturbed by white noise:

$$\Sigma_0: \qquad dx(t) = Ax(t)dt + A_1x(t)dw_1(t) \tag{26.1}$$

(Ito stochastic differential equation). Here $A, A_1 \in \mathbb{R}^{n\times n}$ and w_1 is a standard real-valued Wiener process on a probability space $(\Omega, \mathcal{F}, \mu)$ relative to an increasing family $(\mathcal{F}_t)_{t\in\mathbb{R}_+}$ of σ-algebras $\mathcal{F}_t \subset \mathcal{F}$. For every $x^0 \in \mathbb{R}^n$ there exists a unique solution $x(t) = x(t, x^0)$ of (26.1) on $\mathbb{R}_+ = [0, \infty)$ with $x(0) = x^0$. The system is called *(mean square) stable* if there exists a constant $c > 0$ such that

$$\mathcal{E}\int_0^\infty \|x(t,x^0)\|^2\,dt \le c\|x^0\|^2, \quad x^0 \in \mathbb{R}^n.$$

We will consider both complex and real perturbations of (26.1) and let $\mathbf{K} = \mathbb{C}$ or $\mathbf{K} = \mathbb{R}$. Suppose the system Σ_0 is stable and is perturbed to

$$\Sigma_\Delta: \quad dx(t) = Ax(t)dt + B\Delta(Cx(t))dt + A_1x(t)dw_1(t) + B_1\Delta(C_1x(t))dw_2(t) \tag{26.2}$$

where B, $B_1 \in \mathbb{R}^{n\times\ell}$, C, $C_1 \in \mathbb{R}^{q\times n}$ are given, $\Delta : \mathbf{K}^q \mapsto \mathbf{K}^\ell$ is an unknown Lipschitzian nonlinearity, and w_2 is another standard Wiener process on $(\Omega, \mathcal{F}, \mu)$ which is not necessarily independent of w_1. If $\Delta(y) = \Delta y$ is linear and $w_1 = w_2$ then the perturbed equation (26.2) can be written in the form (26.1) with A replaced by $A + B\Delta C$ and A_1 by $A_1 + B_1\Delta C_1$. Thus our model allows for simultaneous disturbances of both the deterministic and the stochastic parameters of the system. We denote by $\mathcal{N}(\mathbf{K}^q, \mathbf{K}^\ell)$ (resp. $\mathcal{L}(\mathbf{K}^q, \mathbf{K}^\ell)$) the set of Lipschitzian nonlinearities (resp. linear maps) $\Delta : \mathbf{K}^q \mapsto \mathbf{K}^\ell$ with $\Delta(0) = 0$ and define $\|\Delta\|$ to be the minimal $\gamma > 0$ satisfying $\|\Delta(y)\|_{\mathbf{K}^\ell} \leq \gamma\|y\|_{\mathbf{K}^q}$ for all $y \in \mathbf{K}^q$.

Definition: The *stability radius* of Σ_0 with respect to *nonlinear* perturbations of the form $\Sigma_0 \rightsquigarrow \Sigma_\Delta$ is

$$r^w_{\mathbf{K},\mathcal{N}} = \inf\{\|\Delta\|; \Delta \in \mathcal{N}(\mathbf{K}^q, \mathbf{K}^\ell), \Sigma_\Delta \text{ is not stable}\}. \tag{26.3}$$

The stability radius of Σ_0 with respect to *linear* perturbations of the above form is defined analogously and denoted by $r^w_{\mathbf{K},\mathcal{L}}$.

2 Problems

Problem 1: Stability radius. Find characterizations and/or formulas from which the stability radii $r^w_{\mathbf{K},\mathcal{N}}$, $\mathbf{K} = \mathbb{C}$ or $\mathbb{R}$ can be determined. In particular, if $C = C_1$, examine under which conditions $r^w_{\mathbf{K},\mathcal{N}} = \|\mathbf{L}\|^{-1}$ where

$$\mathbf{L} : L^2_w(\mathbb{R}_+; L^2(\Omega, \mathbb{R}^\ell)) \to L^2_w(\mathbb{R}_+; L^2(\Omega, \mathbb{R}^q)), \quad v(\cdot) \mapsto z(\cdot) = Cx(\cdot, 0, v)$$

is the input ouptut operator of the stable system

$$\begin{aligned} dx(t) &= Ax(t)dt + Bv(t)dt + A_1x(t)dw_1(t) + B_1v(t)dw_2(t) \quad (26.4)\\ z(t) &= Cx(t). \end{aligned}$$

It is known that $r^w_{\mathbf{K},\mathcal{N}} \geq \|\mathbf{L}\|^{-1}$. In order to prove equality it is necessary to construct a destabilizing perturbation $\Delta \in \mathcal{N}(\mathbf{K}^q, \mathbf{K}^\ell)$ for (26.2) with norm arbitrarily close to $\|\mathbf{L}\|^{-1}$.

Problem 2: Nonlinear versus linear. Determine whether $r^w_{\mathbb{C},\mathcal{N}} = r^w_{\mathbb{C},\mathcal{L}}$. If $r^w_{\mathbb{C},\mathcal{N}} \neq r^w_{\mathbb{C},\mathcal{L}}$ find characterizations and/or formulas from which $r^w_{\mathbf{K},\mathcal{L}}$, $\mathbf{K} = \mathbb{C}$ or $\mathbb{R}$ can be computed.

A solution of the following problem will be easy once the previous two problems have been solved.

Problem 3: Real versus complex. Determine whether $r^w_{\mathbb{C},\mathcal{N}} = r^w_{\mathbb{R},\mathcal{N}}$ (and $r^w_{\mathbb{C},\mathcal{L}} = r^w_{\mathbb{R},\mathcal{L}}$).

The next problem will be difficult.

Problem 4: Stochastic μ-problem. Solve the stability radius problem for blockdiagonal perturbations of the form

$$A \rightsquigarrow A + \sum_{i=1}^{N} B_i \Delta_i C_i, \quad A_1 \rightsquigarrow A_1 + \sum_{i=1}^{N} B_i^1 \Delta_i C_i^1.$$

and for their nonlinear counterpart.

Remark: Analogous problems can be formulated for other stability concepts (e.g. stability with probability 1, exponential stability in pth mean, $p \geq 1$) and other types of random perturbations (where the Wiener process is replaced by a stationary Gaussian process, coloured noise), see [1].

3 Motivations

The motivation for studying the above problems is pretty obvious since they are just stochastic versions of robust stability problems which have been studied extensively in the deterministic context over the past 15 years. Whilst some results are available for purely stochastic perturbations (i.e. $B = 0$, see the next subsection) it appears that the above problems have hardly been dealt with when *both the drift term and the diffusion term in (26.1)* are disturbed.

The above problems are also of interest in the deterministic context. If $A_1 = 0$ the nominal system (26.1) is deterministic and then the problem is to determine how robust the stability of $\dot{x} = Ax$ is under both deterministic and stochastic parameter perturbations.

4 Available results

It is well known [3] that a system of the form (26.1) is stable iff there exists a matrix $P \prec 0$ such that

$$PA + A^*P + A_1^*PA_1 \succ 0. \tag{26.5}$$

The stabilization of stochastic systems with multiplicative noise has been studied since the late sixties, particularly in the context of linear quadratic optimal control [9]. The subject of robust stability and robust stabilization is of more recent vintage. An early reference is [8] where the problem is considered in an almost disturbance decoupling framework. Formulae for stability radii of stochastic systems have been obtained for various special cases. The first formula was derived for the case $A_1 = 0, B = 0$ in [2]. Morozan [7] extended this formula to the case where $B = 0$ and the nominal system contains a sum of white noise terms. The nonlinear version of

Problem 4 (μ-problem) was solved in [5] for the special case of a deterministic nominal system with purely stochastic blockdiagonal perturbations (i.e. $A_1 = 0$ and $B_i = 0$, $i = 1, \ldots, N$).
All these characterizations of stability radii presuppose that only stochastic parameter disturbances are present ($B = 0$ or $B_i = 0$). In other words, it is assumed that the nominal model is exact in the mean. First results concerning stability radii of systems subjected to simultaneous deterministic and stochastic parameter perturbations have been presented in [4]. A Stochastic Bounded Real Lemma from which – at least in principle – one can determine $\|\mathbf{L}\|$ was proved in [6]. It states that (26.4) is stable with $\|\mathbf{L}\| < \gamma$ iff there exists $P \prec 0$ such that $\gamma^2 I_\ell + B_1^* P B_1 \succ 0$ and

$$PA + A^*P + A_1^*PA_1 - C^*C-$$

$$-(PB + q_{12}A_1^*PB_1)(\gamma^2 I_\ell + B_1^*PB_1)^{-1}(B^*P + q_{12}B_1^*PA_1) \succ 0 \quad (26.6)$$

where the incremental covariance q_{12} is defined by

$$\mathcal{E}((w_1(t) - w_1(s))(w_2(t) - w_2(s))) = q_{12}(t-s), \qquad t, s \in \mathbb{R}_+,\ t > s.$$

Since (26.5) and (26.6) are matrix inequalities without any stochastic features a deterministic analysis of these inequalities seems to be desirable. Surprisingly little work appears to have been done in this direction even for the well known simple and basic inequality (26.5).

Problem 5: Structure theory of (26.5) and (26.6). Find necessary and sufficient algebraic or frequency domain type conditions for the existence of negative definite solutions of (26.5) and (26.6). Develop an algebraic theory of these matrix inequalities and the corresponding matrix equations.

5 References

[1] L. Arnold. Stochastic Differential Equations: Theory and Applications. J. Wiley, New York, 1974.

[2] A. El Bouhtouri and A. J. Pritchard. Stability radii of linear systems with respect to stochastic perturbations. *Systems & Control Letters* 19: 29–33, 1992.

[3] R. Z. Has'minskii. Stochastic stability of differential equations. Sijthoff & Noordhoff, Alphen aan den Rijn, 1980 (translation of the Russian edition, Moscow, Nauka 1969).

[4] D. Hinrichsen and A. J. Pritchard. Stability margins for systems with deterministic and stochastic uncertainty. *Proc. 33rd IEEE Conf. Decision and Control*, Florida 1994.

[5] D. Hinrichsen and A. J. Pritchard. Stability radii of systems with stochastic uncertainty and their optimization by output feedback. *SIAM J. Control and Optimization* 34: 1972-1998, 1996.

[6] D. Hinrichsen and A. J. Pritchard. Stochastic H^∞. IDS-Report 336, University of Bremen, 1996, accepted for publication in *SIAM J. Control and Optimization.*

[7] T. Morozan. Stability radii for some stochastic differential equations. *Stochastics and Stochastics Reports* 54: 281-291, 1995.

[8] J. L. Willems and J. C. Willems. Robust stabilization of uncertain systems. *SIAM J. Control and Optimization* 21: 352–374, 1983.

[9] W. M. Wonham. Optimal stationary control of a linear system with state dependent noise. *SIAM J. Control and Optimization* 5: 486-500, 1967.

27

Monotonicity of performance with respect to its specification in H^∞ control

Hidenori Kimura

Department of Mathematical Engineering and Information Physics
The University of Tokyo
7-3-1, Hongo, Bunkyo-ku
Tokyo 113-8656
JAPAN
kimura@crux.t.u-tokyo.ac.jp

1 H^∞ control problem

Assume that the plant to be controlled is represented by the transfer function matrix $P(s)$, *i.e.*,

$$\begin{bmatrix} z(s) \\ y(s) \end{bmatrix} = P(s) \begin{bmatrix} w(s) \\ u(s) \end{bmatrix} = \begin{bmatrix} P_{11}(s) & P_{12}(s) \\ P_{21}(s) & P_{22}(s) \end{bmatrix} \begin{bmatrix} w(s) \\ u(s) \end{bmatrix}, \tag{27.1}$$

where $z(s)$ denotes an m-dimensional vector of errors to be controlled, $y(s)$ q-dimensional observation vector, $w(s)$ r-dimensional vector of exogeneous signals and $u(s)$ denotes p-dimensional vector of control input. As illustrated in Fig.1, the controller is represented by

$$u(s) = K(s)y(s). \tag{27.2}$$

Then, the closed-loop transfer function from the exogeneous signal $w(s)$ to the error $z(s)$ is given by

$$z(s) = \Phi(s)w(s) \tag{27.3}$$

where $\Phi(s)$ is given by

$$\Phi(s) = P_{11}(s) + P_{12}(s)K(s)(I - P_{22}(s)K(s))^{-1}P_{21}(s). \tag{27.4}$$

If the dependence of $\Phi(s)$ on $K(s)$ needs to be explicitly represented, we write it as $\Phi(s;\ K(s))$.

The H^∞ control problem is formulated as follows:

H^∞ Control Problem:
Given a positive number $\gamma > 0$, find a controller $K(s)$ of (27.2) that stabilizes a closed-loop system of Fig.1 and satisfies

$$\|\Phi(s;\ K(s))\|_\infty < \gamma \tag{27.5}$$

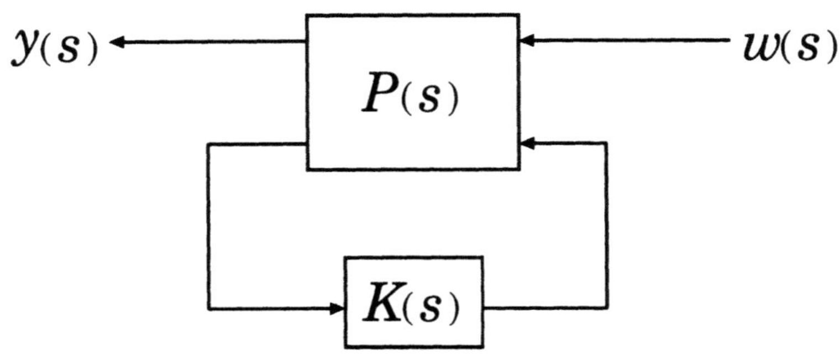

FIGURE 1. Closed-loop system

The problem of H^∞ control was originally proposed by George Zames [1] as an alternative to the optimal regulator. Enormous number of contributions have been made to this problem, and now the H^∞ control is extensively used in many areas of control. Here, we just mention a paper by Doyle, Glover, Khargonekar and Francis [2] which made a breakthrough in this area.

For a special class of plants that satisfy the conditions

$$\text{(A1)} \qquad \text{rank} P_{12}(j\omega) = p, \quad \text{rank} P_{21}(j\omega) = q, \quad \forall\omega$$

a necessary and sufficient condition for the existence of controller that stabilizes the closed-loop system of Fig.1 and satisfies (27.5) is well-known. Under that condition, the controller is given by

$$K(s) = F_{11}(s) + F_{12}(s)U(s)(I - F_{22}(s)U(s))^{-1}F_{21}(s) \tag{27.6}$$

where $F_{ij}(s)$ are represented in terms of the solutions of Riccati equations and $U(s)$ is any stable matrix that satisfies $\|U\|_\infty < 1$. For the detailed description of the controller (27.6), see, e.g., [2]. The controller $K(s)$ that corresponds to $U(s) = 0$ in (27.6), that is $K(s) = F_{11}(s)$, is called the *central controller.* It should be noted that almost all H^∞ controllers that are available by commercial software package are the central controller.

2 Mustafa-Glover conjecture of monotorinity

The specification of the control performance is represented by a scalar $\gamma > 0$, which gives a bound on the H^∞ norm of closed-loop transfer function Φ as given in the inequality (27.5). We call γ a *specification parameter.* Since the specification is required to be satisfied by the strict inequality in (27.5), there is always a gap between the specified upper bound γ and the actual norm of the closed loop transfer matrix $\|\Phi\|_\infty$. Now, we confine our attention only on the central controller. The central controller is then determined by the specification parameter γ. The central controller corresponding to the specification parameter γ is denoted by $K_\gamma(s)$. The actual H^∞ norm of the closed-loop transfer function attained by $K_\gamma(s)$ is a function of γ, which is denoted as

$$h(\gamma) := \|\Phi(s;\ K_\gamma(s))\|_\infty \tag{27.7}$$

The above function represents the actual closed-loop performance that is attained by $K_\gamma(s)$ as a function of the spedification parameter γ which represents our expectation of control performance. The selection of the smaller γ represents expectation for a better performance. Therefore, it is natural to expect that the smaller γ results in a smaller $h(\gamma)$ (better performance). In other words, $h(\gamma)$ must be a monotone function with respect to γ. In a celebrated book [3], Mustafa and Glover conjectured that $h(\gamma)$ is a monotonously increasing function for $\gamma \geq \gamma_{opt}$, where γ_{opt} is the minimum γ for which H^∞ control problem is solvable. A typical profile of $h(\gamma)$ they assumed is illustrated in Fig.2.

Mustafa-Glover Monotonicity Conjecture

$h(\gamma)$ in (27.7) is a monotonously increasing function of γ.

3 A counter example

In [4], Ushida and Kimura considered the conjecture and obtained a simple counter example based on the chain-scattering approach which was extensively developed in [5]. The counter example they showed is a simple

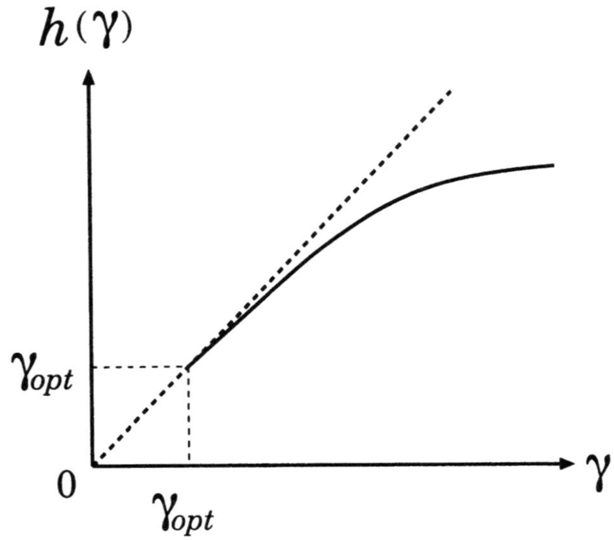

FIGURE 2. A typical profile of $h(\gamma)$

second-order plant with $m = r = p = q = 1$ given by

$$G(s) = D + C(sI - A)^{-1}B \tag{27.8}$$

$$\begin{aligned} A &= \begin{bmatrix} -4 & 25 \\ -10 & 29 \end{bmatrix}, \quad B = \begin{bmatrix} 0.8 & -1 \\ 0.9 & -1 \end{bmatrix} \\ C &= \begin{bmatrix} 10 & -25 \\ 13 & 25 \end{bmatrix}, \quad D = \begin{bmatrix} 0 & 1 \\ 1 & 0 \end{bmatrix}. \end{aligned}$$

This plant actually gives rise the simplest class of H^∞ control problems, namely, the *one-block* problem for which the H^∞ solution is derived by solving two Lyapunov equations, instead of two Riccati equation, for general *four-block* problem.
The γ_{opt} for this plant is calculated to be

$$\gamma_{opt} \simeq 1.\ 134,$$

and it was proven theoretically and numerically in [4] that the function $h(\gamma)$ defined by (27.7) has a local maximum at

$$\gamma = \gamma_m \simeq 1.\ 324,$$

beyond which it decreases up to $\gamma \simeq 1.572$, which is a local minimum.

4 Open problems

Since the plant (27.8) is a counter example to the Mustafa-Glover conjecture, it turns out that in H^∞ control, a stronger performance specification does not necessarily lead to a better performance. Satisfaction of stronger specification may result in worse performance. This might be a crucial drawback of H^∞ control as a design methodology of control systems. Now, the problem arises: *Under what conditions on the plant $P(s)$ the function $h(\gamma)$ given by (27.7) is monotonously increasing with respect to γ ?*
This is indeed a crucial question for the practical applicability of H^∞ control.
In [4], a simple sufficient condition was proven for the monotonicity of $h(\gamma)$ with respect to the number of unstable zeros of $P_{12}(s)$. Ushida and Kimura showed that $h(\gamma)$ is monotonically increasing if $P_{12}(s)$ has at most one unstable zero. In that case, $h(\gamma)$ is explicitly given in the form

$$h(\gamma) = \frac{\beta}{1+\gamma^{-2}\alpha}, \qquad \alpha > 0, \quad \beta > 0.$$

Unfortunately, if $P_{21}(s)$ has more than one unstable zeros, representation of $h(\gamma)$ is prohibitively complex. It should be noted that in the plant (27.8), $P_{12}(s)$ has two unstable zeros at $s = 4$ and $s = 6$.
Another important question is : *Is there a class of H^∞-controllers* (not central controllers) *for which the monotonicity conjecture holds* ?
If we are able to find such a class of controllers, it might provide more preferable controllers from practical point of view than central controllers.

5 Conclusion

We have pointed out an open problem in the area of H^∞ control, which seems to be very important from practical viewpoint. Though the problem seems to be hard, its solution would be expected to give a new insight in the fundamental structure of H^∞ control.

6 References

[1] G.Zames, Feedback and optimal sensitivity: model reference transformations, multiplicative seminorms, and approximate inverses, *IEEE Trans. on Automatic Control,* Vol.AC-26, pp.301-320, 1981.

[2] J.C.Doyle, K.Glover, P.P.Khargonekar and B.A. Francis, State-space solutions to standard H_∞ and H_2 control problems, *IEEE Trans. on Automatic Control,* Vol.AC-34, pp.831-847, 1989.

[3] D.Mustafa and K.Glover, *Minimum Entropy H_∞ Control,* Springer Verlag, 1990.

[4] S.Ushida and H.Kimura, A counter example to Mustafa-Glovers' monotonicity conjecture, *Systems and Control Letters,* Vol.28, pp.129-137, 1996.

[5] H.Kimura, *Chain-Scattering Approach to H^∞-Control,* Birkhauser, 1996.

28

Stable estimates in equation error identification: An open problem

Roberto López-Valcarce*[1] and Soura Dasgupta**[2]

The University of Iowa
Iowa City, IA 52242
USA
dasgupta@hitchcock.eng.uiowa.edu

1 The problem

Identification algorithms generally fall into two distinct categories: output error and equation error. The difference between them is best exemplified by figures 1 and 2 respectively,

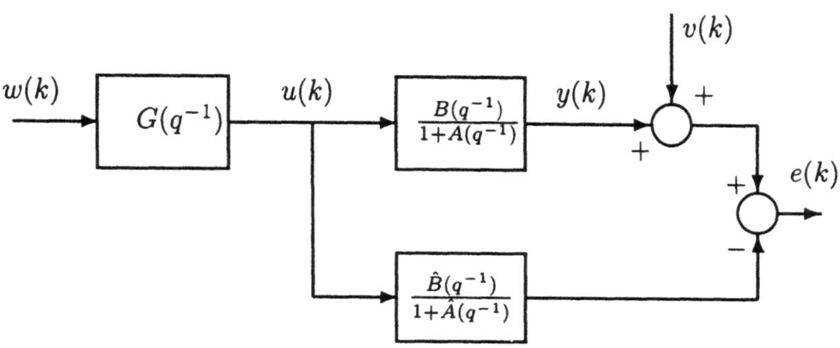

FIGURE 1. Output error configuration

where the unknown plant to be identified has transfer function

$$H(q^{-1}) = \frac{B(q^{-1})}{1 + A(q^{-1})} = \frac{\sum_{i=0}^{m} b_i q^{-i}}{1 + \sum_{j=1}^{n} a_j q^{-j}} \tag{28.1}$$

with q the forward shift operator. Assume that the plant input $u(k)$ is the output of a coloring filter $G(q^{-1})$, in turn excited by zero mean, wide sense

[1]Supported by Fundación Pedro Barrié de la Maza under grant no. 340056.
[2]Supported in part by NSF grant ECS-9350346.

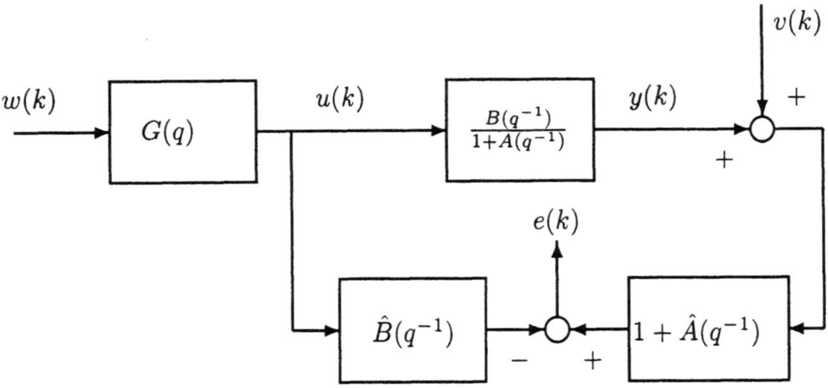

FIGURE 2. Equation error configuration

stationary (WSS), unit variance white noise. We will assume that the plant output $y(k)$ is corrupted by zero mean WSS white noise $v(k)$. In the sequel, define

$$\hat{B}(q^{-1}) = \sum_{i=0}^{m} \hat{b}_i q^{-i}, \quad \hat{A}(q^{-1}) = \sum_{j=1}^{n} \hat{a}_j q^{-j} \tag{28.2}$$

which serve as the respective estimates of $B(q^{-1})$ and $A(q^{-1})$. For each algorithm one obtains $\hat{B}(q^{-1})$ and $\hat{A}(q^{-1})$ to minimize the variance of $e(k)$ in the pertinent figure.

Where as the underlying optimization problem in figure 2 is convex, the one in figure 1 is not. Consequently, while adaptive algorithms approximating the gradient descent optimization of the setting in figure 2 are globally convergent, the corresponding algorithm for figure 1 is only locally convergent [3]. At the same time, while the global optimum in figure 1 represents unbiased estimates of $B(q^{-1})$ and $A(q^{-1})$, the values of $\hat{B}(q^{-1})$ and $\hat{A}(q^{-1})$ corresponding to the global optimum in figure 2 will in general be biased. It is possible in fact [1] for this bias to result in an $\hat{A}(q^{-1})$ for which $1/(1+\hat{A}(q^{-1}))$ is unstable, even when $1/(1+A(q^{-1}))$ is stable. Such instability is clearly undesirable in both signal processing and control. Accordingly, it is important to determine conditions that ensure that $\hat{A}(q^{-1})$ represents a stable plant estimate.

Accordingly the open problem we pose is as follows:

Problem: Consider the equation error setting depicted in figure 2, with $w(k)$ and $v(k)$, zero mean and white and $1/(1 + A(q^{-1}))$ stable. Characterize the class of $B(q)/(1 + A(q^{-1}))$ for which the $\hat{B}(q)/(1 + \hat{A}(q^{-1}))$ obtained by minimizing the variance of $e(k)$ is stable.

2 Known facts

Very large and very small Signal to Noise Ratios at the output always lead to stable estimates, [1]. Should $n = m$ and $G(q^{-1})$ be autoregressive with order n, then again stability of the estimate is assured, [2].
Suppose $\hat{B}(q^{-1}) = B(q^{-1}) = 1$. Then the following additional facts are known, [4]. With $A(q^{-1})$ as in (28.2) we will denote the vector $\mathbf{a} = [a_1, a_2, \ldots, a_n]^T$. We will call $\mathbf{a}$ stable if $1/(1 + A(q^{-1}))$ is stable. We say $\mathbf{a} \in \mathcal{S}$ if $\mathbf{a}$ is stable, and under $B = \hat{B} = 1$, for all stable $G(q^{-1})$ in figure 2, the $1 + \hat{A}(q^{-1})$ that minimizes the variance of $e(k)$ has a stable inverse.
In other words, in the all-pole case, equation error identification always yields a stable plant estimate whenever $\mathbf{a} \in \mathcal{S}$. Then we have the following theorem:

Theorem 3 *With $\mathcal{S}$ as above, $\mathbf{a} \in \mathcal{S}$ iff for all positive definite symmetric Toeplitz $n \times n$ matrix $\mathbf{R}$, $\mathbf{R}(\mathbf{I} + \mathbf{R})^{-1}\mathbf{a}$ is stable.*

A complete characterization of $\mathcal{S}$ also remains an open problem. In the case $n = 2$, $\mathcal{S}$ corresponds to a subset of the celebrated stability triangle. In particular in a_1, a_2 space this region is a pentagon with corner coordinates: $(0, -1)$, $(-1.5, .5)$, $(-1, 1)$, $(1.5, .5)$, $(1, 1)$.

A sufficient condition for the convergence of a certain adaptive output error algorithms is that $1 + A(q^{-1})$ be strictly positive real (SPR), [3] i.e. it is stable and $\text{Re}[1 + A(e^{-j\omega})] > 0\ \forall\omega$. It is thus of interest to see if a similar SPR condition features in the characterization of S. Indeed, for n=2, $1 + A(q^{-1})$ SPR implies that $\mathbf{a} \in \mathcal{S}$. However $\mathbf{a} \in \mathcal{S}$ does not imply $1 + A(q^{-1})$ SPR. It is then natural to ask if the same holds in higher dimensions. The answer to this question is no. Indeed take the example $\mathbf{a} = [.05, -.9, -.1]$. Observe $1 + .05q^{-1} - .9q^{-2} - .1q^{-3}$ is SPR. Yet with R the positive definite Toeplitz matrix

$$\begin{bmatrix} 6.778 & 2.7217 & -4.4444 \\ 2.7217 & 6.778 & 2.7217 \\ -4.4444 & 2.7217 & 6.778 \end{bmatrix},$$

$\mathbf{R}(\mathbf{I} + \mathbf{R})^{-1}\mathbf{a}$ is unstable.
We end with the following conjecture. The set S is convex. This is true for n=2 and n=3.

3 References

[1] T. Soderstrom and P. Stoica, "On the stability of dynamic models obtained by least squares identification," *IEEE Trans. Auto. Contr.*, vol. AC-26, pp. 575-577, 1981.

[2] P. A. Regalia, "An unbiased equation error identifier and reduced order appriximations", *IEEE Trans. on Signal Processing,* vol. 42, pp. 1397-1412, 1994.

[3] C. R. Johnson, Jr., *Lectures in Adaptive Parameter Estimation*, Prentice Hall, 1988.

[4] Roberto López-Valcarce and Soura Dasgupta, "Stable Estimates in Equation Error Identification", in *Proceedings of CDC*, Dec, 1997.

29
Elimination of latent variables in real differential algebraic systems

Iven Mareels* and Jan C. Willems**

*Department of Electrical and Electronic Engineering
University of Melbourne
Parkville, Victoria 3032
AUSTRALIA
iven@mullian.mu.oz.au

**Mathematics Institute
University of Groningen
P.O. Box 800
9700 AV Groningen
THE NETHERLANDS
willems@math.rug.nl

1 Introduction

The purpose of this note is to attract attention to what is called the *elimination problem*, which we believe to be an important but largely open problem in mathematical modelling. A typical way of modelling dynamical systems is by the method of hierarchical *tearing* and *zooming:* a system is decomposed into subsystems until a level is reached where the subcomponents have mathematical models that are assumed to be "known". This procedure is described for example in [17]. Electrical circuits and robotic kinematic chains form the prototype examples of this sort of first principles modelling. These ideas also lie at the basis of modelling concepts as bond-graphs and object-oriented computer-assisted procedures used frequently for instance in chemical process modelling.

As a consequence of the introduction of internal interconnections, the resulting model invariably contains more variables than those that the model aims at describing. We call these additional variables *latent variables*, in order to distinguish them from the *manifest variables*, as we call the variables of primary interest in the modelling process. The problem discussed in this note is the extent to which the latent variables can be eliminated.

We now formulate this question in mathematical terms for differential algebraic systems. Let f be a vector of polynomials. We will be interested in polynomials over both the real and complex field, since the results will be quite different. Let $\mathbb{K} = \mathbb{R}$ or $\mathbb{C}$. The vector of polynomials f is used to define a system of differential equations in the variables w and ℓ. The w's are the manifest variables, and the ℓ's are the latent variables, and the fact that they both appear is of the essence. The order of this differential equation is denoted by n. The assumption that each of the differential equations has the same order n in both w and ℓ can be achieved by putting the appropriate coefficients to zero. This is reflected in the following notation for the variables in f, the indeterminates:

$$w, w^{(1)}, w^{(2)}, \cdots, w^{(n)}, \ell, \ell^{(1)}, \cdots, \ell^{(n)} \tag{29.1}$$

where each of the $w^{(k)}$'s consist of q variables, and each of the $\ell^{(k)}$'s consists of d variables. This leads to

$$f(w, w^{(1)}, \cdots, w^{(n)}, \ell, \ell^{(1)}, \cdots, \ell^{(n)}), \tag{29.2}$$

a vector of polynomials with coefficients in $\mathbb{K}$ in $(n+1)(q+d)$ indeterminates. Expression (29.2) leads to the system of differential equations

$$f \circ (w, \frac{d}{dt}w, \cdots, \frac{d^n}{dt^n}w, \ell, \frac{d}{dt}\ell, \cdots, \frac{d^n}{dt}\ell) = 0. \tag{29.3}$$

The end result of the tearing and zooming modelling process referred to above is typically such a system of differential equations, assuming that each of the subsystems are described by polynomial differential expressions. Of course, often more complicated functions than polynomials appear, but for the purposes of this note, we concentrate on polynomial functions. Note that static (i.e., algebraic) equations can be accommodated in (29.3) by having differential equations of order zero. This is important, since first principles modelling invariably leads, because of the interconnection equations, to a predominance of algebraic as compared to dynamic equations.

2 Elimination of latent variables

Equation (29.3) defines a *dynamical system with latent variables* as this term is defined in [18]:

$$\Sigma_f = (\mathbb{R}, \mathbb{K}^q, \mathbb{K}^d, \mathcal{B}_f) \tag{29.4}$$

where $\mathbb{R}$ is the time-axis, $\mathbb{K}^q$ the space of manifest variables, $\mathbb{K}^d$ the space of latent variables, and the *full behavior* $\mathcal{B}_f$ is defined as

$$\mathcal{B}_f = \{(w, \ell) : \mathbb{R} \to \mathbb{K}^q \times \mathbb{K}^d) | \text{ (29.3) is satisfied}\} \tag{29.5}$$

We are deliberate vague about the solution concept that is involved in the definition of $\mathcal{B}_f$, since flexibility in this respect will be required in order to obtain satisfactory results. When we suppress the latent variables in (29.4), we obtain the *manifest system* induced by (29.4), defined as

$$\Sigma = (\mathbb{R}, \mathbb{K}^q, \mathcal{B}) \tag{29.6}$$

with *manifest behavior*

$$\mathcal{B} = \{w : \mathbb{R} \to \mathbb{K}^q \mid \exists\, \ell : \mathbb{R} \to \mathbb{K}^d \text{ such that } (w, \ell) \in \mathcal{B}_f\} \tag{29.7}$$

We view (29.6) as a system in which the latent variables have been eliminated.
The question which we address is:

What sort of equations, formulae, describe the manifest behavior $\mathcal{B}$?

We call the problem of obtaining such formulae the *elimination problem.* In order to obtain a clean theory, it may be necessary to amend the definitions of $\mathcal{B}_f$ and $\mathcal{B}$ suitably, e.g., by allowing solutions defined on finite intervals only, or by allowing solutions that are distributions or hyper-functions.

2.1 The linear case

When f is a linear map (29.3) leads to the system of differential equations

$$R(\frac{d}{dt})w = M(\frac{d}{dt})\ell \tag{29.8}$$

with R and M polynomial matrices over $\mathbb{K}$. In this case it is known [10] that after elimination of the latent variables we obtain again a system of linear differential equations. More precisely, for any pair of polynomial matrices R and M over $\mathbb{K}$, there exist a polynomial matrix R' over $\mathbb{K}$ such that the set

$$\{w \in C^\infty(\mathbb{R}, \mathbb{K}^q) \mid \exists\, \ell \in C^\infty(\mathbb{R}, \mathbb{K}^d) \text{ such that (29.8) holds}\} \tag{29.9}$$

consists *exactly* of all C^∞ solutions of

$$R'(\frac{d}{dt})w = 0 \tag{29.10}$$

It is important to note that, in order to cope rigorously with lack of controllability and common factors, this elimination result requires more than just transfer function thinking. This elimination result has recently been generalized to constant coefficient PDE's [9] and to time-varying systems [8].

2.2 The complex case

A more complicated situation has been studied in [3] for complex differential algebraic systems, i.e., with f a vector of polynomials and $\mathbb{K} = \mathbb{C}$. In this case, elimination (using an algebraically motivated solution concept for algebraic differential equations) is shown to lead to a system of resultant equations, i.e., it is shown that $\mathcal{B}$ leads to the finite set of $\mathcal{B}_\alpha$'s (with $\alpha \in A$, and A a finite index set), each described by combined algebraic differential equations and inequations. The relation between $\mathcal{B}$ and the $\mathcal{B}_\alpha$'s appears to be that $\mathcal{B}$ is contained in $\bigwedge_{\alpha \in A} \mathcal{B}_\alpha$, where $\bigwedge_{\alpha \in A} \mathcal{B}_\alpha$ denotes the *concatenation product* of the $\mathcal{B}_\alpha$'s.

Each of the $\mathcal{B}_\alpha$'s is specified by a finite set of polynomials

$$f_\alpha(w, w^{(1)}, \cdots, w^{(n_\alpha)}) \tag{29.11}$$

over $\mathbb{C}$ and one polynomial

$$g_\alpha(w, w^{(1)}, \cdots, w^{(n_\alpha)}) \tag{29.12}$$

over $\mathbb{C}$ such that

$$\begin{aligned} \mathcal{B}_\alpha = \{w : \mathbb{R} \to \mathbb{C}^q \quad | \quad & f_\alpha \circ (w, \frac{d}{dt}w, \cdots, \frac{d^{n_\alpha}}{dt^{n_\alpha}}w) = 0 \\ & (g_\alpha \circ (w, \frac{d}{dt}w, \cdots, \frac{d^{n_\alpha}}{dt^{n_\alpha}}w))(t) \neq 0 \text{ for all } t \in \mathbb{R}\} \end{aligned} \tag{29.13}$$

In fact, [3] also describes an algorithm for passing from f to (f_α, g_α)'s. In addition, this result remains true in the more general case in which the starting point (29.3) contains also inequations. As such, it is much more logical to start in the complex case from the very beginning with a system of differential algebraic equations *and* differential algebraic inequations (or a finite union of these). And indeed, that is what is done in [3]. Note, however, that in general $\mathcal{B}$ will be properly contained in $\bigwedge_{\alpha \in A} \mathcal{B}_\alpha$. Moreover, in the dynamic case, when trajectories come from differential equations, and pass from one of the $\mathcal{B}_\alpha$'s to another, certain *gluing conditions* have to be satisfied. We will return to these in the next section.

3 Open problems

The question raised is to study the case that $\mathbb{K} = \mathbb{R}$. It is easy to see that in addition to inequations also inequalities are needed in this case. A concrete conjecture therefore is that the resulting equations (29.7) will then consist of a finite union of systems of differential equations, differential inequations, and differential inequalities. Hence each of the resulting $\mathcal{B}_\alpha$'s will be specified by f_α's, g_α's, and h_α's, with the f_α's and g_α's as (29.11, 29.12), but with real coefficients, and

$$h_\alpha(w, w^{(1)}, \cdots, w^{(n_\alpha)}) \tag{29.14}$$

a system of polynomials over $\mathbb{R}$, such that

$$\begin{aligned} \mathcal{B}_\alpha = \{w : \mathbb{R} \to \mathbb{R}^q \quad | \quad & f_\alpha \circ (w, \frac{d}{dt}w, \cdots, \frac{d^{n_\alpha}}{dt^{n_\alpha}}w) = 0 \\ & (g_\alpha \circ (w, \frac{d}{dt}w, \cdots \frac{d^{n_\alpha}}{dt^{n_\alpha}}w))(t) \neq 0 \text{ for all } t \in \mathbb{R} \\ & (h_\alpha \circ (w, \frac{d}{dt}w, \cdots \frac{d^{n_\alpha}}{dt^{n_\alpha}}w))(t) \geq 0 \text{ for all } t \in \mathbb{R}\} \end{aligned} \tag{29.15}$$

and $\mathcal{B} \subset \bigwedge_{\alpha \in A} \mathcal{B}_\alpha$.

Of course, once this result has been established, it follows that in the real case, it is more natural to start, instead of with (29.3), with a vector of differential algebraic equations, inequations, and inequalities (or a finite union of these), and take the development from there. An excellent reference that gives a starting point for the problem put forward here is [5].

The $\mathcal{B}_\alpha$'s contain the manifest behavior $\mathcal{B}$, in the sense that $\mathcal{B} \subset \bigwedge_{\alpha \in A} \mathcal{B}_\alpha$. The question occurs what has to be added in order to specify $\mathcal{B}$ exactly. The concatenations cannot occur freely, in the sense that for example the concatenation of $w_1 \in \mathcal{B}_{\alpha_1}$ and $w_2 \in \mathcal{B}_{\alpha_2}$ at $t = 0$ will be an element of $\mathcal{B}$ only if certain conditions matching the derivatives of w_1 at $t = 0^-$ with those of w_2 at $t = 0^+$ are satisfied. These relations are called the *gluing conditions.*

In conclusion, the open problem consists of the following parts.

(i) Establish the resulting f_α's, g_α's, h_α's, in the real case (in particular, prove that there are only a finite number of these).

(ii) Prove (both in the complex and in the real case) that each $w \in \mathcal{B}$ is the concatenation of trajectories from the $\mathcal{B}_\alpha$'s.

(iii) Establish (both in the complex and in the real case) the gluing conditions.

The important system theoretic implication of all this is that nonlinear differential equations form not a particularly natural starting point for the manifest behavior of a nonlinear dynamical system. Since each system is in some sense the result of interconnecting subsystems, it is unclear how the problem of elimination of latent (interconnection) variables can be avoided. Inequalities and inequations will be introduced, not necessarily because of "hard" constraints that may be present in a system, but because they are introduced during the process of elimination.

The problems presented here are natural in the development of the study of algebraic difference or differential equations. Such systems where introduced in the system theory literature by Sontag in his thesis (see [12]), and further developed in [7, 14] (see [15] for a tutorial exposition). Other system theory work in this areas centers around the names of Fliess [6, 7] and Glad [4, 5]. It is interesting to observe that the gluing conditions can been seen as additional motivation for hybrid systems work as reported for example in [1, 16].

We remark in closing that the motivation of this open problem comes from an awareness of the urgency to generalize the algorithmic aspects of the behavioral theory (elimination, controllability, observability, control, etc., etc.) from linear systems to the differential algebraic systems. Differential algebraic systems have been studied from a control perspective for instance by Fliess and Glad [6, 7]. It is likely that Ritt's algorithm [5, 4, 11] and Gröbner bases techniques [3] can be used effectively in the problems proposed.

4 References

[1] R.W. Brockett, Hybrid models for motion control systems, pages 29-53 of *Essays on Control: Perspectives in the Theory and Its Applications*, edited by H.L. Trentelman and J.C. Willems, Birkhäuser, 1993.

[2] D. Cox, J. Little, and D. O'Shea, *Ideals, Varieties and Algorithms: An Introduction to Computational Algebraic Geometry and Commutative Algebra*, 2nd edition, Springer Verlag, 1997.

[3] S. Diop, Elimination in Control Theory, *Mathematics of Control, Signals, and Systems,* volume 4, pages 17-32, 1991.

[4] S.T. Glad, Implementing Ritt's algorithm of differential algebra, pages 610-614 of *Proceedings of the IFAC Symposium NOLCOS'92*, Bordeaux, 1992.

[5] S.T. Glad, Solvability of differential algebraic equations and inequalities: an algorithm, *Proceedings of the 1997 European Control Conference*, Brussels, 1997.

[6] M. Fliess, Automatique et corps différentiels, *Forum Mathematicae*, pages 227-238, 1989.

[7] M. Fliess and S.T. Glad, An algebraic approach to linear and nonlinear control, pages 223-267 of *Essays on Control: Perspectives in the Theory*

and Its Applications, edited by H.L. Trentelman and J.C. Willems, Birkhäuser, 1993.

[8] S. Fröhler and U. Oberst, Continuous time-varying linear systems, manuscript, 1998.

[9] H.K. Pillai and S. Shankar, A behavioural approach to control of distributed systems, *SIAM Journal on Control and Optimization*, to appear.

[10] J.W. Polderman and J.C. Willems, *Introduction to Mathematical System Theory: A Behavioral Approach*, Springer-Verlag, 1998.

[11] J.F. Ritt, *Differential Algebra*, AMS, 1950.

[12] E.D. Sontag, *Polynomial Response Maps*, Springer-Verlag, 1979.

[13] Y. Wang and E.D. Sontag, Algebraic differential equations and rational control systems, *SIAM Journal on Control and Optimization*, volume 30, pages 1126-1149, 1992.

[14] Y. Wang and E.D. Sontag, Orders of input/output differential equations and state space dimensions, *SIAM Journal on Control and Optimization*, volume 33, pages 1102-1127, 1995.

[15] E.D. Sontag, Spaces of observables in nonlinear control, in *Proceedings of the International Congress of Mathematicians 1994*, volume 2, Birkäuser, pages 1532-1545, 1995.

[16] A.J. van der Schaft, J.M. Schumacher, Complementarity modeling of hybrid systems, to appear in *IEEE Transactions on Automatic Control*, volume 43, 1998.

[17] J.C. Willems, The behavioral approach to systems and control, *Journal of the Society of Instrument and Control Engineers of Japan*, volume 34, pages 603-612, 1995. Appeared also in the *European Journal of Control* volume 2, pages 250-259, 1996.

[18] J.C. Willems, Paradigms and puzzles in the theory of dynamical systems, *IEEE Transactions on Automatic Control*, volume 36, pages 259-294, 1991.

30
How conservative is the circle criterion?

Alexandre Megretski

Laboratory for Information and Decision Systems
Massachusetts Institute of Technology
Cambridge, MA 02139
USA
ameg@mit.edu

1 Problems description

The following notation and terminology will be used. By a *stable transfer function* we mean a rational scalar transfer function $G = G(z)$ with real coefficients and no poles in the area $|z| \geq 1$. $\|G\|_\infty$ denotes the so-called *H-infinity norm* of G, defined as

$$\|G\|_\infty = \max_{|z|=1} |G(z)|.$$

For a stable transfer function $G = G(z)$, T_G denotes the LTI system with transfer function G. Mathematically, T_G can be viewed as the operator $T_G : l_2 \mapsto l_2$, where l_2 is the standard Hilbert space of one-sided sequences of real numbers. T_G is then defined by $V(z) = G(z)W(z)$ for $v = T_G w$, where $V(z)$ and $W(z)$ are the Z-transforms of v and w respectively. In other words, T_G is a frequency-domain multiplication operator.
For a bounded sequence $\delta = \delta[k]$ of real numbers, M_δ denotes the time-varying system of "modulation by δ". Mathematically, M_δ is represented by the operator of multiplication by δ in the time domain, $M_\delta : \ l_2 \mapsto l_2$, where $w = M_\delta v$ means $w[k] = \delta[k]v[k]$ for all k.) We will be interested in relating the maximal values of G and δ to *stability* of feedback interconnections of T_G and M_δ. Mathematically, we say that the interconnection of T_G and M_δ is *stable* if the operator $I - T_G M_\delta$ is invertible.

Problem 1. Does there exist a finite constant $\gamma > 0$ with the following feature: for any stable transfer function $G = G(z)$ with $\|G\|_\infty \geq \gamma$ there exists a sequence $\delta = \delta[k] \in [-1, 1]$ such that the feedback interconnection of T_G and M_δ is not stable.

Problem 2. Does there exist a finite constant $\gamma > 0$ with the following feature: for any $\theta \in [0, \pi]$ and for any polynomial $b(z) = b_2 z^2 + b_1 z + b_0$ such that $|b(e^{j\theta})| \geq \gamma$, there is no single causal linear time-invariant feedback $u = K(s)y$ which stabilizes the system

$$y = \frac{1}{z^2 + 2\cos(\theta)z + 1}[z^2 \;\; b(z)] \begin{bmatrix} u \\ M_\delta u \end{bmatrix} \tag{30.1}$$

for *any* real sequence $\delta = \delta[k] \in [-1, 1]$.

Problem 3. Does there exist a finite constant $\gamma > 0$ with the following feature: for any cyclic n-by-n real matrix

$$H = \begin{bmatrix} h_0 & h_1 & \dots & h_n \\ h_n & h_0 & \dots & h_{n-1} \\ & & \vdots & \\ h_1 & h_2 & \dots & h_0 \end{bmatrix} \tag{30.2}$$

such that $\sigma_{\max}(H) \geq \gamma$, there exists a non-zero real vector x such that $|y_i| \geq |x_i|$ for all $i = 0, 1, ..., n$, where

$$x = \begin{bmatrix} x_0 \\ x_1 \\ \vdots \\ x_n \end{bmatrix}, \; y = \begin{bmatrix} y_0 \\ y_1 \\ \vdots \\ y_n \end{bmatrix} = Hx.$$

2 Motivation

Problems 1-3 (in case the answer is "yes" in all of them) establish an unexpected link between control theory, harmonic analysis, and combinatorics. In harmonic analysis, Problem 1 represents an attempt to establish an important connection between the time-domain and frequency domain multiplications. In control theory, Problem 1 is a natural follow-up question to the statement of the famous *circle criterion*, which states that the feedback interconnection of T_G and M_δ is stable when $\|G\|_\infty < 1$ and $\max_k |\delta[k]| \leq 1$. It is known that the circle criterion is conservative (i.e. the feedback interconnection can be stable for all such δ even when $\|G\|_\infty > 1$), but just *how conservative could it be*?
Problem 2 discusses the possibility of robust stabilization of a second-order uncertain system using a linear and time-invariant controller. It is easy to show that a nonlinear state feedback can stabilize system (30.1) for any b. It was also shown in [1] that a static linear state feedback cannot stabilize a continuous-time analog of system (30.1) when $|b(e^{j\omega})|$ is large enough. It

is intuitively appealing to conjecture that using a dynamic LTI feedback cannot stabilize the system either. Also, it can be easily seen that Problems 1 and 2 are equivalent.
Problem 3, which is a "finite-dimensional" version of Problem 1, is also a typical example of a conjectured finiteness of the gap between the minimum in a specially structured non-convex quadratic optimization problem (finding the maximal ρ such that $|y_i| \geq \gamma |x_i|$ for some $y = Hx$, $x \neq 0$, where H is a given cyclic matrix), and its natural convex relaxation (finding the maximal singular value of H). It is easy to see that a positive answer to Problem 3 implies positive answers to Problems 1 and 2. Most likely, Problems 1 and 2 are equivalent to Problem 3.

3 References

[1] F. Blanchini, and A. Megretski "Robust state feedback control of LTV systems: nonlinear better than linear", *1997 ECC*, also: to appear in *IEEE Trans. Autom. Ctrl.*.

31

On chaotic observer design

Henk Nijmeijer
Systems, Signals and Control Department
Faculty of Mathematical Sciences
University of Twente
P.O. Box 217
7500 AE Enschede
THE NETHERLANDS
h.nijmeijer@math.utwente.nl
and
Faculty of Mechanical Engineering
Eindhoven University of Technology
P.O. Box 513
5600 MB Eindhoven
THE NETHERLANDS

1 Problem formulation

Chaos synchronization is an extremely exciting subject within the physics and communications society and deals with the remarkable observation that certain chaotic dynamic systems exhibit asymptotically identical motion under weak coupling between the systems. Mathematically, synchronization of two systems – sometimes called transmitter and receiver – can be cast in the following form. Given the single output system Σ_1 on $\mathbb{R}^n$ defined by

$$\Sigma_1 : \begin{cases} \dot{x} &= f(x) = f(x_1, x_2, \cdots, x_n) \\ y &= h(x) = x_1 \end{cases}$$

introduce the system Σ_2 as a copy of Σ_1 with the first state component identical to y, that is

$$\Sigma_2 : \begin{cases} \dot{\tilde{x}} &= f(x_1, \tilde{x}_2, \cdots, \tilde{x}_n) \\ y &= h(\tilde{x}) = \tilde{x}_1 \end{cases}$$

Then the transmitter Σ_1 (with transmitted signal x_1) and receiver Σ_2 synchronize, that is to say, for arbitrary initial conditions of Σ_1 and Σ_2, we have

$$\lim_{t\to\infty} \mid x(t) - \tilde{x}(t) \mid = 0.$$

Synchronization as defined above can be viewed as an observer problem in that Σ_2 acts as an observer for Σ_1. Clearly, the choice of Σ_2 as an observer is not general, but works well in some specific examples of chaotic systems, cf. [1]. Usually, a full order observer for Σ_1 is given as a system Σ_3 described by

$$\Sigma_3 : \begin{cases} \dot{\tilde{x}} &= \tilde{f}(\tilde{x}, y) \\ \tilde{y} &= h(\tilde{x}) \end{cases}$$

with $\tilde{x} \in \mathbb{R}^n$ and $\tilde{f}$ a smooth vector field parametrized by the measurements y, and Σ_3 is chosen such that $e(t) = x(t) - \tilde{x}(t)$ asymptotically converges to zero as $t \to \infty$ for all initial conditions x_0 and $\tilde{x}_0$, and moreover, if $e(t_0) = 0$, then $e(t) = 0$ for all $t \geq t_0$, see e.g. [2] and references inthere for a discussion on observers. Clearly the system Σ_2 forms a very particular choice for an observer for Σ_1, and it is therefore natural to seek, in case Σ_2 does not act as an observer, other possible choices for an observer. A rather natural attempt consists of selecting as a candidate observer for Σ_1 a system given by

$$\Sigma_4 : \begin{cases} \dot{\tilde{x}} &= f(\tilde{x}) + k(y - \tilde{y}) \\ \tilde{y} &= h(\tilde{x}) = \tilde{x}_1 \end{cases}$$

where k is constant $(n, 1)$-vector. In the latter case typical high gain observers – with large entries in k – are included in the candidate observer Σ_4 (see [4] for cases where such high gain observer works).

Within the synchronization literature (cf. [1]) a popular choice for a candidate successful receiver is given by a system of the form Σ_4 with as extra requirement that the vector k has all entries zero except the first one. In this case the candidate observer exploits ideas from sliding mode control in that the first component of e is forced to converge (fast) to zero. As an example we consider, as in [1], the synchronization of two chaotic Chua circuits:

$$C_1 : \begin{cases} \dot{x}_1 &= \alpha(x_2 - x_1 - f(x_1)) \\ \dot{x}_2 &= x_1 - x_2 + x_3 \\ \dot{x}_3 &= -\beta x_2 \\ y &= x_2 \end{cases}$$

$$C_2 : \begin{cases} \dot{\tilde{x}}_1 &= \alpha(\tilde{x}_2 - \tilde{x}_1 - f(\tilde{x}_1)) \\ \dot{\tilde{x}}_2 &= \tilde{x}_1 - \tilde{x}_2 + \tilde{x}_3 + k_2(x_2 - \tilde{x}_2) \\ \dot{\tilde{x}}_3 &= -\beta \tilde{x}_2 \\ \tilde{y} &= \tilde{x}_2 \end{cases}$$

Here $f(x_1) = \gamma x_1 + \delta(\mid x_1 + 1 \mid - \mid x_1 - 1 \mid)$, and $\alpha, \beta, \gamma, \delta$ specific constants, $k_2 > 0$ is the design parameter. Note that C_1 exhibits chaotic motion for

certain parameter choices for α, β, γ and δ. In [1] it is conjectured, on the basis of simulations and experiments that for certain values of k_2 the two systems C_1 and C_2 synchronize.
The problem we are interested in is whether indeed C_2 acts as an observer for C_1. The answer to this question is in fact negative; namely if we initialize C_1 at the origin, then C_2 will only converge towards C_1 if it is initialized in 0. But the question is whether the system C_2 synchronizes with C_1 in case the initial condition for C_1 is different from 0, no matter what the initial condition $\tilde{x}_0$ for C_2 is. Simulations as well as experimental results from [1] suggest that the synchronization of C_2 with C_1 indeed takes place, though it is clear that $x_0 = 0$ (together with its stable manifold) forms a measure zero exception.

2 Background and motivation

Synchronization of systems has been observed in various systems, but seldom a link is made with the observer problem from control (cf. [2]). Given the work as described in [1], and references in there, it is fair to say that there is a need for studying the nonlinear observer problem. Although this problem in its full generality is notably difficult, there are several classes of systems for which the problem might be feasible. In particular chaos synchronization, advocated as an extremely important and powerful tool in secure communications (cf. [1]), leads to a question as proposed before. As an intriging subproblem one might wonder what role is played by the chaos in the dynamics; usually in observer design or synchronization the nature of the dynamics of the system Σ_1 should not be decisive in error convergence property. It is for that reason that one may think that, as the synchronization of C_1 and C_2 at least suggests in simulation and practice, the chaotic nature of the Chua-circuit C_1 is crucial. Other ideas about exploiting the chaotic nature of the system Σ_1 can be found in [5]. It is worth mentioning that the given problem formulation dealing with a pair of Chua circuits is one out of many. Other illustrations of synchronized motions occur for instance in pairs of parametrically driven pendulums, see [3].

3 References

[1] T. Kapitaniak, *Controlling chaos*, Academic Press, London, 1996.

[2] H. Nijmeijer and I.M.Y. Mareels, "An observer looks at chaos", IEEE Trans. Circuits and Systems I, Vol. 44, pp. 882-890, 1997.

[3] E.J. Banning, *On the dynamics of two coupled parametrically driven pendulums*, PhD thesis, University of Twente, 1997.

[4] J. Gauthier, H. Hammouri and S. Othman, "A simple observer for nonlinear systems, applications to bioreactors", IEEE Trans. Automat. Contr. Vol. 37, pp. 875-880, 1992.

[5] H. Nijmeijer, "On synchronization of chaotic systems", Proceedings 36th IEEE Conference on Decision and Control, San Diego, Cal., pp. 384-388, 1997.

32
The minimal realization problem in the max-plus algebra

Geert Jan Olsder* and Bart De Schutter**

*Department of Mathematics and Informatics
Delft University of Technology
2600 GA Delft
THE NETHERLANDS
g.j.olsder@math.tudelft.nl

**ESAT-SISTA
K. U. Leuven
3001 Leuven
BELGIUM
Bart.DeSchutter@esat.kuleuven.ac.be

1 Description of the problem

Given an arbitrary real sequence $\{g_i\}_{i=1}^{\infty}$ elegant necessary and sufficiency conditions are known for the existence of an $n \times n$ matrix A, an $n \times 1$ vector b and a $1 \times n$ vector c, for some appropriate n, such that

$$g_i = cA^{i-1}b \qquad \text{for } i = 1, 2, \ldots \tag{32.1}$$

The elements of A, b and c are supposed to be real numbers. An additional requirement might be that n, which determines the sizes of A, b and c, must be as small as possible. In that case n is called the *minimal system order* and the triple A, b and c is a *minimal realization*. Efficient algorithms to calculate a minimal realization are known (see, e.g., [11]).
The problem considered in this chapter arises when the underlying algebra is the so-called max-plus algebra [2, 3] rather than the conventional algebra tacitly used above. One obtains the max-plus algebra from the conventional algebra by replacing addition by maximization and multiplication by addition. These operations are indicated by $\oplus$ (maximization) and $\otimes$ (addition). In the max-plus algebra one for instance has

$$\begin{pmatrix} 1 & 4 \\ -3 & 0 \end{pmatrix} \otimes \begin{pmatrix} 5 \\ 1 \end{pmatrix} = \begin{pmatrix} (1 \otimes 5) \oplus (4 \otimes 1) \\ (-3 \otimes 5) \oplus (0 \otimes 1) \end{pmatrix} = \begin{pmatrix} 6 \\ 2 \end{pmatrix}.$$

2 Motivation

In conventional system theory the sequence $\{g_i\}_{i=1}^{\infty}$ arises as the impulse response of the linear, finite-dimensional, discrete-time, time-invariant SISO[1] state space description

$$x(k+1) = Ax(k) + bu(k) \ , \quad y(k) = cx(k),$$

see (32.1). The problem considered here is to compute a minimal realization and to characterize the minimal system order for max-plus linear systems, i.e., systems of the form

$$x(k+1) = A \otimes x(k) \oplus b \otimes u(k) \ , \quad y(k) = c \otimes x(k) \ . \tag{32.2}$$

In spite of its misleading simple formulation, this problem has met with formidable difficulties.

3 History and partial results

3.1 Characterization of max-plus-algebraic impulse responses

A necessary and sufficient condition for a sequence $\{g_i\}_{i=1}^{\infty}$ to be the impulse response of a system that can be described by a model of the form (32.2) is that the sequence is *ultimately periodic* [8, 9], i.e.,

$$\begin{aligned} &\exists m, \lambda_0, \ldots, \lambda_{m-1}, k_0 \text{ such that } \forall k \geq k_0 : \\ &\qquad g_{km+m+s} = \lambda_s^{\otimes^m} \otimes g_{km+s} \quad \text{for } s = 0, 1, \ldots, m-1 \ . \end{aligned}$$

where $\lambda^{\otimes^m} = \lambda \times m$. An ultimately periodic sequence can be decomposed into subsequences (one for each λ_s), each of which shows, after a transient part, a periodic behaviour superimposed upon a linear drift (the drift is characterized by the parameter λ_s).

3.2 The minimal system order

We define so-called Hankel matrix $H(\alpha, \beta)$ of size $\alpha \times \beta$ as

$$H(\alpha,\beta) = \begin{pmatrix} g_1 & g_2 & \cdots & g_\beta \\ g_2 & g_3 & \cdots & g_{\beta+1} \\ \vdots & \vdots & \ddots & \vdots \\ g_\alpha & g_{\alpha+1} & \cdots & g_{\alpha+\beta-1} \end{pmatrix} . \tag{32.3}$$

---

[1]SISO: single-input single-output. Generalizations to multiple-input multiple-output systems exist, but will not be emphasized here.

In conventional system theory the minimal system order is given by the rank of the Hankel matrix $H(\infty,\infty)$. However, in contrast to linear algebra the different notions of rank (like column rank, row rank, minor rank, ...) are in general not equivalent in the max-plus algebra[2].
Let $H = H(\infty,\infty)$. It can be shown [8] that the minimal system order is equal to the smallest integer r for which there exist an $\infty \times r$ matrix U, an $r \times \infty$ matrix V and an $r \times r$ matrix A such that $H = U \otimes V$ and $U \otimes A = \overline{U}$, where $\overline{U}$ is the matrix obtained by removing the first row of U.
The different notions of matrix rank in the max-plus-algebra can be used to obtain lower and upper bounds for the minimal system order. The so-called max-plus-algebraic minor rank and the Schein rank of H provide lower bounds [8, 9]. At present, there are no efficient (i.e., polynomial time) algorithms to compute the max-plus-algebraic minor rank or the Schein rank of a matrix. The max-plus-algebraic weak column rank of H provides an upper bound [8, 9]. Efficient methods to compute this rank are described in [3, 8].

3.3 Minimal state space realization: partial results

Transformation to conventional algebra

There exists a transformation from the max-plus algebra to the linear algebra that is based on the following equivalences:

$$x \oplus y = z \quad \Leftrightarrow \quad e^{xs} + e^{ys} \sim ce^{zs}\ ,\ s \to \infty \tag{32.4}$$

$$x \otimes y = z \quad \Leftrightarrow \quad e^{xs} \cdot e^{ys} = e^{zs} \quad \text{for all } s > 0 \tag{32.5}$$

with $c = 2$ if $x = y$ and $c = 1$ otherwise.
Using this transformation the minimal realization problem in the max-plus algebra can be mapped to a minimal realization problem for matrices with exponentials as entries and with conventional addition and multiplication as basic operations [12, 13]. This implies that we can use the techniques from conventional realization theory to obtain a minimal realization and afterwards (try to) transform the results back to the max-plus algebra. However, only realizations with positive coefficients for the leading exponentials can be mapped back to the max-plus algebra, and it is not always obvious how and whether such a realization can be constructed.

Partial state space realization

The *partial* minimal realization problem is defined as follows: given a finite sequence $g_1, g_2, \ldots, g_N$, find A, b and c such that $g_i = c \otimes A^{\otimes^{i-1}} \otimes b$ for

[2] An overview of the relations between the different ranks in the max-plus algebra can found on p. 122 of [8].

$i = 1, 2, \ldots, N$. It can be shown that this leads to a system of so-called max-plus-algebraic polynomial equations and that such a system can be recast as an Extended Linear Complementarity Problem (ELCP) [5, 6]. This enables us to solve the partial minimal realization problem and by applying some limit arguments this results in a realization of the entire impulse response. However, it can be shown that the general ELCP is NP-hard.

Special sequences of Markov parameters

For some special cases, e.g., if the sequence $\{g_i\}_{i=1}^{\infty}$ exhibits uniformly up-terrace behavior [16, 17], or if the sequence exhibits a convex transient behavior and a so-called ultimately geometric behavior with period 1 [4, 10], there exist methods to efficiently compute minimal state space realizations.

4 Related fields

Based on the relations (32.4) and (32.5) it is easy to verify that there exists a connection between the minimal realization problem in the max-plus algebra and the minimal realization problem for nonnegative systems. Indeed, some of the results obtained in system theory for nonnegative systems also hold in the max-plus algebra (see, e.g., [7]). For more information on the minimal realization problem for nonnegative systems the reader is referred to [1, 15].

Remark: For a more extended overview of known results, open problems and additional references in connection with the minimal realization problem in the max-plus algebra the interested reader is referred to [14].

Acknowledgments

Bart De Schutter is a senior research assistant with the F.W.O. (Fund for Scientific Research-Flanders). This research was sponsored by the Flemish Government (GOA-MIPS), the Belgian Federal Government (IUAP P4-02, IUAP P4-24), and the European Commission (TMR (Alapedes), network contract ERBFMRXCT960074). The part of Geert Jan Olsder was also sponsored by the same EU contract.

5 References

[1] B.D.O. Anderson, M. Deistler, L. Farina, and L. Benvenuti, "Nonnegative realization of a linear system with nonnegative impulse response," *IEEE Transactions on Circuits and Systems*, vol. 43, pp. 134–142, 1996.

[2] F. Baccelli, G. Cohen, G.J. Olsder, and J.P. Quadrat, *Synchronization and Linearity.* New York: John Wiley & Sons, 1992.

[3] R.A. Cunninghame-Green, *Minimax Algebra*, vol. 166 of *Lecture Notes in Economics and Mathematical Systems.* Berlin, Germany: Springer-Verlag, 1979.

[4] R.A. Cunninghame-Green and P. Butkovič, "Discrete-event dynamic systems: The strictly convex case," *Annals of Operations Research*, vol. 57, pp. 45–63, 1995.

[5] B. De Schutter, *Max-Algebraic System Theory for Discrete Event Systems.* PhD thesis, Faculty of Applied Sciences, K.U.Leuven, Leuven, Belgium, 1996.

[6] B. De Schutter and B. De Moor, "Minimal realization in the max algebra is an extended linear complementarity problem," *Systems & Control Letters*, vol. 25, no. 2, pp. 103–111, May 1995.

[7] B. De Schutter and B. De Moor, "Matrix factorization and minimal state space realization in the max-plus algebra," in *Proceedings of the 1997 American Control Conference (ACC'97)*, Albuquerque, New Mexico, USA, pp. 3136–3140, June 1997.

[8] S. Gaubert, *Théorie des Systèmes Linéaires dans les Dioïdes.* PhD thesis, Ecole Nationale Supérieure des Mines de Paris, France, July 1992.

[9] S. Gaubert, "On rational series in one variable over certain dioids," Tech. rep. 2162, INRIA, Le Chesnay, France, Jan. 1994.

[10] S. Gaubert, P. Butkovič, and R. Cunninghame-Green, "Minimal $(\max, +)$ realization of convex sequences", *SIAM Journal on Control and Optimization*, vol. 36, no. 1, pp. 137-147, Jan. 1998.

[11] T. Kailath, *Linear Systems.* Englewood Cliffs, New Jersey: Prentice-Hall, 1980.

[12] G.J. Olsder, "Some results on the minimal realization of discrete-event dynamic systems," Tech. rep. 85-35, Delft University of Technology, Faculty of Technical Mathematics and Informatics, Delft, The Netherlands, 1985.

[13] G.J. Olsder, "On the characteristic equation and minimal realizations for discrete-event dynamic systems," in *Proceedings of the 7th International Conference on Analysis and Optimization of Systems*, Antibes, France, vol. 83 of *Lecture Notes in Control and Information Sciences*, pp. 189–201, Berlin, Germany: Springer-Verlag, 1986.

[14] G.J. Olsder, B. De Schutter and R.E. de Vries, "The minimal state space realization problem in the max-plus algebra: An overview", Tech. rep. 97-107, ESAT-SISTA, K.U.Leuven, Leuven, Belgium, Dec. 1997.

[15] J.M. van den Hof, *System Theory and System Identification of Compartmental Systems.* PhD thesis, Faculty of Mathematics and Natural Sciences, University of Groningen, Groningen, The Netherlands, Nov. 1996.

[16] L. Wang and X. Xu, "On minimal realization of SISO DEDS over max algebra," in *Proceedings of the 2nd European Control Conference*, Groningen, The Netherlands, pp. 535–540, June 1993.

[17] L. Wang, X. Xu, and R.A. Cunninghame-Green, "Realization of a class of discrete event sequence over max-algebra," in *Proceedings of the 1995 American Control Conference*, Seattle, Washington, pp. 3146–3150, June 1995.

33

Input design for worst-case identification

Jonathan R. Partington

School of Mathematics
University of Leeds
Leeds, LS2 9JT
UNITED KINGDOM
J.R.Partington@leeds.ac.uk

1 Description of the problem

We are concerned with the worst-case identification of a linear discrete-time time-invariant system $y = h * u + v$, i.e.,

$$y(t) = \sum_{k=0}^{t} h(k)u(t-k) + v(t), \qquad t = 0, 1, 2, \ldots,$$

where h is a BIBO stable impulse response (that is, $\sum |h(k)| < \infty$), u is an input sequence that we can choose, y is an observed output sequence, and v is a small perturbation, which may be due either to the fact that the true system is not genuinely linear or to the effects of measurement inaccuracy. We shall suppose that u is bounded, indeed we may suppose that $|u(t)| \leq 1$ for all t.

The object of such an experiment is to use data $y(0), \ldots, y(T-1)$ to construct a model $\hat{h}$ for h, which is close in the sense that $\|\hat{h} - h\|$ is small whenever T is sufficiently large and the perturbations v are sufficiently small.

Let us write $H(z)$ for the analytic function $\sum_{k=0}^{\infty} h(k)z^k$. It is of interest to consider three distinct norms:

$$\|h\|_1 = \sum_{k=1}^{\infty} |h(k)|, \qquad \|H\|_\infty = \sup_{|z|<1} |H(z)|, \qquad \text{and}$$

$$\|h\|_2 = \|H\|_2 = \left(\sum_{k=0}^{\infty} |h(k)|^2 \right)^{1/2}.$$

It is well-known that $\|h\|_2 \leq \|H\|_\infty \leq \|h\|_1$, so that it is hardest to obtain good approximations in the ℓ_1 norm. All three norms are commonly seen in problems of systems and control.
The general question that we pose here is: *What is a good choice for the input sequence u?* Although these questions are posed in a non-probabilistic framework, their solutions enrich the classical theory of identification (cf. [7]), as well.

2 Motivation and History

With the development of robust control, it has been considered desirable to produce identification algorithms that can yield approximate models, with guaranteed error bounds under suitable hypotheses. It has also been worth investigating the extent to which functional analytical techniques are able to shed light on classical questions such as those of input design.
Most of the work on control-oriented identification has been published since 1991 (early references include [6] for frequency-domain identification and [8] for the time-domain version). The problems we describe can also be phrased quite naturally in the language of Information-Based Complexity (see, e.g. [11] for details). Other related asymptotic issues, including related some results on nonlinear systems, have been treated in [2]. Further reading on much of the work in this area can be found in the survey article [9] and the monograph [10].

3 Results and specific questions

For identification in the presence of bounded errors, it has been found desirable to provide input sequences which extract the system from the error in an efficient manner in the simple case when the system h is a FIR model, that is $h(t) = 0$ for $t \geq n$. To do this, we require, for a fixed $\delta > 0$,

$$\max_{0 \leq t \leq N+n-2} |(h * u)(t)| \geq \delta \|h\| \qquad \text{for all such } h.$$

We call the least such N the *sample complexity*. More sophisticated error models, and more sophisticated model sets for h, can also be considered: for example, a large class of rational models was considered in [4]. The question remains the same—given an n-dimensional subspace X_n, and $\delta > 0$, what is the least N for which there is an input u with $\|u\|_\infty \leq 1$, such that the above inequality holds for all $h \in X_n$?
For the norms above, and FIR models, the following designs are appropriate:

1. The ℓ_2 case. A good design would be to construct a sequence

$$u(0), \ldots, u(N-1),$$

with $N = O(n)$ such that (for some positive δ independent of N),

$$\inf_{|z|=1} |u(0) + u(1)z + \ldots + u(N-1)z^{N-1}| \geq \delta\sqrt{N}$$

(see [10]). It is not known whether this can be done for real u. For complex u, it can [1], and a real input design based on such a complex u can be made with some loss of efficiency.

2. The H_∞ case. Here a design with $N = O(n^2)$, based on nth roots of unity, has optimal sample complexity [5].

3. The ℓ_1 case. Galois sequences—suitable sequences of ± 1 (see [8]) of length N exponential in n—are known to be essentially optimal for $\delta = 1$ (see [3]). For smaller δ, what easily-computed (shorter) designs are available? Are there any n-dimensional subspaces of ℓ_1 (i.e., not FIR models) for which the sample complexity is sub-exponential in n?

It would be an interesting challenge to combine these designs in some way, so that one obtained a design that was efficient for both H_∞ and ℓ_2—which would need to be defined in some precise sense.
Finally, suppose that we require a slowly-varying input (to avoid component wear in process control). What reasonably efficient designs are possible?

4 References

[1] Benke, G. (1992). On the minimum modulus of trigonometric polynomials. *Proc. Amer. Math. Soc.*, **114**, 757–761.

[2] Dahleh, M.A., Sontag, E.D., Tse, D.N.C., and Tsitsiklis, J.N. (1995). Worst-case identification of nonlinear fading memory systems. *Automatica*, **31**, 503–508.

[3] Dahleh, M.A., Theodosopoulos, T., and Tsitsiklis, J.N. (1993). The sample complexity of worst-case identification of F.I.R. linear systems. *Systems Control Lett.*, **20**, 157–166.

[4] Harrison, K.J., Partington, J.R., and Ward, J.A. (1997). Complexity of identification of linear systems with rational transfer functions, preprint.

[5] Harrison, K.J., Ward, J.A., and Gamble, D.K. (1996). Sample complexity of worst-case H^∞-identification. *Systems Control Lett.*, **27**, 255–260.

[6] Helmicki, A.J., Jacobson, C.A., and Nett, C.N. (1991). Control oriented system identification: a worst-case/ deterministic approach in H^∞. *IEEE Trans. Automat. Control*, **36**, 1163–1176.

[7] Ljung, L. (1987). *System identification.* Prentice-Hall, Englewood Cliffs, New Jersey.

[8] Mäkilä, P.M. (1991). Robust identification and Galois sequences. *Int. J. Control*, **54**, 1189–1200.

[9] Mäkilä, P.M., Partington, J.R., and Gustafsson, T.K. (1995). Worst-case control-relevant identification. *Automatica*, **31**, 1799–1819.

[10] Partington, J.R. (1997). *Interpolation, identification and sampling.* Oxford University Press.

[11] Traub, J.F., Wasilkowski, G.W., and Wozniakowski, H. (1988). *Information-based complexity.* Academic Press, London.

34

Max-plus-times linear systems

Max Plus[1]
INRIA-Rocquencourt
Domaine de Voluceau BP105
F-78153 Le Chesnay Cedex
FRANCE

1 Description of the problem

Let A and B be (m,n) matrices with real nonnegative entries. Let C and D be (p,n) matrices with entries in $\overline{\mathbb{R}} = \mathbb{R} \cup \{-\infty\}$. We denote by $\otimes$ the max-plus matrix product defined by

$$[E \otimes F]_{ij} = \max_k (E_{ik} + F_{kj}) \ .$$

Let δ denote the backward shift operator on sequences $x = (x_k)_{k \in \mathbb{Z}}$ with entries in $\overline{\mathbb{R}}$, defined by $(\delta x)_k = x_{k-1}$. Let $A(\delta)$,$B(\delta)$, [resp. $C(\delta)$ and $D(\delta)$] be matrices whose entries are monomials [resp. max-plus monomials] in δ with nonnegative coefficients [resp. with coefficients in $\overline{\mathbb{R}}$].
We are interested in solving the following problems.

1. Describe the set of n-vectors X with entries in $\overline{\mathbb{R}}$ satisfying

$$\text{(I)} \begin{cases} AX = BX \ , \\ C \otimes X = D \otimes X \ . \end{cases}$$

   In the first equation we adopt the convention $0 \times (-\infty) = 0$.

2. Describe the set of n-vectors of sequences X satisfying

$$\text{(II)} \begin{cases} A(\delta)X = B(\delta)X \ , \\ C(\delta) \otimes X = D(\delta) \otimes X \ . \end{cases}$$

3. Describe the set of couples (λ, X), where X is an n-vector with entries in $\overline{\mathbb{R}}$ and $\lambda \in \mathbb{R}$, satisfying

$$\text{(III)} \begin{cases} A(\lambda)X = B(\lambda)X \ , \\ C(\lambda) \otimes X = D(\lambda) \otimes X \ , \end{cases}$$

[1]The Max Plus working group currently consists of: Marianne Akian, Guy Cohen, Stéphane Gaubert, Jean-Pierre Quadrat and Michel Viot.

where $A_{ij}(\lambda)$, $B_{ij}(\lambda)$ denote the standard evaluations of the corresponding monomials, and $C_{ij}(\lambda)$, $D_{ij}(\lambda)$ denote the max-plus evaluations of the corresponding monomials (the evaluation of a max-plus monomial $m(\delta) = a\delta^n$ at λ (a real number) is defined by $m(\lambda) = n\lambda + a$).

2 Motivations

Such problems arise in at least two different contexts.

1. *Markov Decision processes.* Classical stochastic dynamic programming equations correspond to the second problem (II). Indeed we can partition the vector X into (Y, Z). Then, choosing the matrices $A(\delta) = (I, 0)$, $B(\delta) = (0, B')$, $C(\delta) = (\epsilon, E)$, $D(\delta) = (\delta D', \epsilon)$ (where I is the standard identity matrix, E the max-plus identity matrix and ϵ the zero max-plus matrix), System (II) describes the recurrence
$$\begin{cases} Y_k = B'Z_k \,, \\ Z_k = D' \otimes Y_{k-1} \,. \end{cases}$$
If we are interested in the component Z we obtain
$$Z_k = D' \otimes (B'Z_{k-1}) \,,$$
which is a standard stochastic dynamic programming equation as soon as $B'1 = 1$. The asymptotics of these problems when n goes to ∞ leads to Problem (III). Indeed, the equation
$$Z = D' \otimes (B'Z) + \lambda \,,$$
is a standard stochastic dynamic programming equation for computing the maximal cost by unit of time in the ergodic case [18].

2. *Simulation of general Petri nets.* The dynamic of a general Petri net can be described by special classes of the second type of equations (see [14] Th.II.2), which are more general than the stochastic dynamic programming equations. For some particular routing policies, simulating Petri nets is equivalent to solving stochastic dynamic programming equations (see [4]).

3 Available results

Clearly a lot of results are known in particular cases, but the general theory does not exist.

1. When C and D are max-plus zero matrices, we are in the standard linear algebraic situation.

2. When A and B are conventional zero matrices, we are in the max-plus linear situation.

   (a) When C is the max-plus identity matrix, Problem (II) corresponds to deterministic dynamic programming.

   (b) When the matrix D has only one max-plus nonzero column, Problem (I) can be solved using residuation theory (see for example [2], [1, Ch.4.]).

   (c) A Cramer theory exists for Problem (I) with general C and D matrices (see [1, Ch.3 Sect.4],[10, Ch.3],[15]). This problem can also be solved by elimination methods [3, 11],[10, Ch.3].

   The references [5, 6] may be useful to understand the kernels and the images of max-plus linear operators. See also [17, 12] for available results on semimodules and semirings.

3. Some special instances of Problem (I) are seen in [9, Ch.3 and Ch.4] as extended linear complementary problems. The set of solutions, which is an union of faces of polyedra, cannot be simple in full generality. A kind of max-plus algebraic geometry has to be developped for solving this problem for matrices with integer entries. Some preliminary results on max-plus polynomials can be found in [1, Ch.3 Sect.6],[8, Sec. VIII].

4. Pure standard algebra or max-plus eigenvalues problems are understood, see [7, 13, 16, 10, 1] for the max-plus case. The Markov decision process case is also standard [18]. The problem with simultaneous dependence, in δ, of A in one hand, and C and D in the other hand, is not homogeneous and may have no practical interest. For example, in the stochastic dynamic programming case, B and A do not depend of δ.

4 References

[1] F. Baccelli, G. Cohen, G.J. Olsder, and J.P. Quadrat : "Synchronization and Linearity", Wiley, (1992).

[2] T.S. Blyth and M.F. Janowitz: " Residuation Theory", Pergamon press, (1972).

[3] P. Butkovic and G Hegedüs : "An elimination method for finding all solutions of the system of linear equations over an extremal algebra", Ekonomicko-matematicky Obzor, 20, (1984).

[4] G. Cohen, S. Gaubert and J.P. Quadrat : "Asymptotic Throughput of Continuous Timed Petri Nets" Proceedings of IEEE-CDC Conference (December 1995).

[5] G. Cohen, S. Gaubert and J.P. Quadrat : "Kernels, Images and Projections in Dioids" Proceedings of WODES'96, Edinburgh, (1996).

[6] G. Cohen, S. Gaubert and J.P. Quadrat : "Linear Projector in the Max-plus Algebra" IEEE Mediterraneen Conference on Control, Cyprus, (July 1997).

[7] R. Cunninghame-Green : "Minimax Algebra", L.N.166 on Economics and Math. Systems, Springer Verlag, (1979).

[8] R. Cunninghame-Green : "Minimax Algebra and Applications", Advances in Imaging and Electron Physics, 90, (1995).

[9] B. De Schutter : "Max-Algebraic system Theory for Discrete Event Systems", Thesis Dissertation, Leuven University, (February 1996).

[10] S. Gaubert : "Théorie des systèmes linéaires dans les dioïdes", Thesis dissertation, École des Mines de Paris, (July 1992).

[11] S. Gaubert and M. Plus "Methods and Applications of Max-plus Linear Algebra" Proceedings of STACS'97, Springer LNCS 1200,(1997)

[12] J.S. Golan : "The theory of semirings with applications in mathematics and theoretical computer science" volume 54, Longman Sci & Tech., (1992).

[13] M. Gondran and M. Minoux :"L'indépendance linéaire dans les dioides", Bul. DER, série C, (1) p.67-90, EDF Clamart France (1978).

[14] L. Libeaut : "Sur l'utilisation des dioïdes pour la commande des systèmes événements discrets", Thesis Dissertation École Centrale de Nantes, (September 1996).

[15] M. Plus : Linear systems in (max, +)-algebra. "Proceedings of the 29th Conference on Decision and Control", Honolulu, (Dec. 1990).

[16] V.P. Maslov and S.N. Samborskii : "Idempotent Analysis", AMS (1992).

[17] E. Wagneur : " Moduloids and Pseudomodules 1. dimension theory", Disc. Math. (98): 57-73, (1991).

[18] P. Whittle : "Optimization over Time" Vol.1 and 2 Wiley, (1982 and 1983).

[19] U. Zimmerman : "Linear and Combinatorial Optimization in Ordered Algebraic-Structures", North Holland, (1981).

35
Closed-loop identification and self-tuning

Jan Willem Polderman

Department of Applied Mathematics
University of Twente
P.O.Box 217
7500 AE Enschede
THE NETHERLANDS
J.W.Polderman@math.utwente.nl

1 Introduction and problem description

Consider the discrete-time SISO system

$$\begin{aligned} y(k+n) + a_{n-1}y(k+n-1) + \cdots + a_0 y(k) = \\ b_{n-1}u(k+n-1) + \cdots + b_0 u(k). \end{aligned} \tag{35.1}$$

A dynamic controller for (35.1) is given by:

$$\begin{aligned} u(k+n-1) + c_{n-2}u(k+n-2) + \cdots + c_0 u(k) = \\ d_{n-1}y(k+n-1) + d_{n-2}y(k+n-2) + \cdots + d_0 y(k). \end{aligned} \tag{35.2}$$

In polynomial notation (35.1,35.2) become

$$A(\sigma)y = B(\sigma)u \qquad C(\sigma)u = D(\sigma)y, \tag{35.3}$$

where $(\sigma y)(k) = y(k+1)$ and the polynomials $A(\xi), B(\xi), C(\xi), D(\xi)$ are defined in the obvious way. Of course, the controller polynomials $(C(\xi), D(\xi))$ depend, via the control objective, on the system polynomials $(A(\xi), B(\xi))$. We denote the mapping that assigns the controller parameters to the system parameters by f. We refer to f as the controller design. The problem addressed in this paper is to classify all controller designs f that posses the *self-tuning property* which we define below. This problem is relevant in indirect adaptive control where the system parameters are unknown and are estimated on the basis of the closed-loop behavior. Instrumental in the analysis of adaptive control systems is the *equilibrium equation*.

This equation describes the set of systems that are closed-loop indistinguishable from the true system. To be more precise, let the true system be defined by $(A^0(\xi), B^0(\xi))$ and let the controller be designed on the basis of an estimate $(A(\xi), B(\xi))$, i.e., $(C(\xi), D(\xi)) = f(A(\xi), B(\xi))$. We call $(A(\xi), B(\xi))$ closed-loop indistinguishable from $(A^0(\xi), B^0(\xi))$ if the (autonomous) behaviors defined by $(A^0(\xi), B^0(\xi), f(A(\xi), B(\xi)))$ and $(A(\xi), B(\xi), f(A(\xi), B(\xi)))$ are the same, i.e., for every input-output pair (u, y):

$$A^0(\sigma)y = B^0(\sigma)u,\ C(\sigma)u = D(\sigma)y \Leftrightarrow A(\sigma)y = B(\sigma)u,\ C(\sigma)u = D(\sigma)y, \tag{35.4}$$

where $(C(\xi), D(\xi)) = f(A(\xi), B(\xi))$. Note that in general the notion of closed-loop indistinguishability is not symmetric. We say that the controller design f has the *self-tuning property* if f is constant on the set of systems that are closed-loop indistinguishable from $(A^0(\xi), B^0(\xi))$ for every $(A^0(\xi), B^0(\xi))$ for which f is defined.

2 History and motivation

Before we discuss the mathematical nature of the problem we shall briefly comment on the motivation and the history of the problem.

In indirect, certainty equivalence based adaptive control systems, the controller is chosen on the basis of the current estimate as if it represents the true system. The estimates are obtained from the input-output measurements. If the estimate is closed-loop indistinguishable from the true system, then the input-output data do not falsify the estimate and hence it is no longer updated.

In general, the controller designed on the basis of an estimate differs from the desired controller (the one based on the true system), however, if f possesses the self-tuning property then every closed-loop indistinguishable system yields the same controller (since f is constant on every set of closed-loop indistinguishable systems) and since the true system is trivially closed-loop indistinguishable, this controller is actually the desired controller.

If the closed-loop adaptive system is not externally excited then the best we can obtain from the identification process is a system that is closed-loop indistinguishable from the true system and it is a fact [1] that in this case the set of closed-loop indistinguishable systems contains an infinite number of systems, so not just the true system. It should be clear that it is beneficial that f has the self-tuning property. It has been proved in [2, 1] that if the control objective is pole-assignment, then the controller design has the self-tuning property. A controller design that does not posses the self-tuning property is Linear Quadratic control, see [2, Chapter II.3].

It can be argued that control objectives in terms of the closed-loop behavior only (such as pole-assignment), are likely to have the self-tuning property.

Likewise, control objectives that are based on some sort of optimization of the closed-loop behavior and where a trade-off between the objective function and the system behavior is apparent (such as LQ control) can be expected not to have the self-tuning property.
In [3, 4] an attempt is made to classify all controller designs that have the self-tuning property. The approach taken there is an extension of a solution for the first-order case that was given in [2].
Next we describe the problem in a self-contained mathematical fashion. We first treat the first-order case and subsequently the general case.

3 The first order case

In the first order case the system/controller equations become:

$$y(k+1) + ay(k) = bu(k) \qquad u(k) = dy(k) \tag{35.5}$$

An example of a controller design that has the self-tuning property is pole-assignment. Let the desired closed-loop pole be denoted by α, then the controller as a function of (a, b) is given by $d = f(a, b) = \frac{\alpha+a}{b}$. A pair (a, b) is closed-loop indistinguishable from a given pair (a^0, b^0) if and only if

$$-a^0 + b^0 f(a, b) = -a + bf(a, b). \tag{35.6}$$

Equation (35.6) is the equilibrium equation for $n = 1$. It is easy to see that if (a, b) satisfies (35.6), then $f(a, b) = f(a^0, b^0)$ so that indeed f has the self-tuning property.
Hence, for the first order case the problem is: Classify all functions f for which for every $(a^0, b^0), (a, b)$ the equation (35.6) implies $f(a, b) = f(a^0, b^0)$. The above example shows that $f(a, b) = \frac{\alpha+a}{b}$ provides a whole class of such functions. Notice that these functions are not defined for $b \neq 0$. This is not unreasonable from a control theoretic point of view, since $b = 0$ corresponds to noncontrollable systems. In [3, Theorem 4.1] the following theorem is proved.
Theorem. *(n=1) Let $f : \mathbb{R} \times \mathbb{R} \setminus \{0\} \rightarrow \mathbb{R}$ have the self-tuning property. Assume that there exist (a_1, b_1) and (a_2, b_2) such that:*

$$f(a_1, b_1) \neq f(a_2, b_2), \tag{35.7}$$

then there exists an $\alpha \in \mathbb{R}$, such that for all $(a, b) \in \mathbb{R} \times \mathbb{R} \setminus \{0\}$

$$f(a, b) = \frac{\alpha + a}{b}. \tag{35.8}$$

If we would drop the assumption (35.7), then every constant f would also yield a solution. This assumption, however, is automatically implied if we require the controller design to be stabilizing. Weird solutions are enabled if we do not care too much about the set on which f is defined. Two examples of such functions are

$$f(a,b) = \frac{\alpha + a}{\beta + b} \qquad f(a,b) = \frac{1}{4}\left[b + \sqrt{-8a + b^2}\right]. \tag{35.9}$$

4 The higher order case

The general case ($n \geq 1$) may be approached using either the original input/output framework or a state space representation thereof. In the input-output representation, using the behavioral framework [5], it can be seen that a pair $(A(\xi), B(\xi))$ is closed-loop indistinguishable from $(A^0(\xi), B^0(\xi))$ if and only if there exists a unimodular matrix $U(\xi)$ such that

$$U(\xi)\begin{bmatrix} A^0(\xi) & -B^0(\xi) \\ D(\xi) & -C(\xi) \end{bmatrix} = \begin{bmatrix} A(\xi) & -B(\xi) \\ D(\xi) & -C(\xi) \end{bmatrix}, \tag{35.10}$$

where of course $(C(\xi), D(\xi)) = f(A(\xi), B(\xi))$. It can be shown that (35.10) is equivalent to a set of equations in the coefficients of the respective polynomials. These equations form the equilibrium equation for the higher order case. To get the equilibrium equation in an explicit form we represent the i/o system in a non-minimal state space system with $\psi(k) = \text{col}(y(k), \ldots, y(k-n+1), u(k-1), \cdots, u(k-n+1))$ as the $2n-1$-dimensional state. The resulting state space equations are

$$\psi(k+1) = F\psi(k) + gu(k). \tag{35.11}$$

More details about these state space equations may be found in [1]. In terms of the systems matrices F and g, the equilibrium equation becomes

$$F^0 + g^0 h(F, g) = F + gh(F, g) \tag{35.12}$$

where $h(F, g)$ denotes the state feedback controller based on (F, g). The problem in terms of the state space representation now becomes: determine all functions h that posses the self-tuning property, i.e., for which (35.12) implies $h(F, g) = h(F^0, g^0)$. It is not difficult to see that the i/o formulation and the state space formulation of the problem are equivalent.
To rule out solutions like (35.9) for the first-order case, it seems reasonable to add at least two assumptions. Firstly, f should be defined on the set of pairs of polynomials that correspond to controllable systems, i.e., pairs that are co-prime. Equivalently h should be defined on the set of all controllable pairs (F, g). Secondly, the trivial solution where f (h) is constant is avoided by requiring that the controller design yields asymptotically stable systems.
As for the first order case, pole-assignment yields a family of solutions.
Theorem. *Let $P(\xi)$ be a monic polynomial of degree $2n - 1$.*

(a). Let f be such that for $(C(\xi), D(\xi)) = f(A(\xi), B(\xi))$ there holds $A(\xi)$ $C(\xi) - B(\xi)D(\xi) = P(\xi)$. For all co-prime pairs $(A^0(\xi), B^0(\xi))$ and $(A(\xi), B(\xi))$ for which there exists unimodular $U(\xi)$ such that (35.10) holds, we have that $f(A(\xi), B(\xi)) = f(A^0(\xi), B^0(\xi))$.
(b). Let h be such that for all controllable pairs (F, g) the characteristic polynomial of $F + gh(F, g)$ equals $P(\xi)$, then for all controllable pairs (F^0, g^0), (F, g) for which (35.12) holds, we have that $h(F, g) = h(F^0, g^0)$.

Proof: See [3, 4, Theorem 3.1] or [1, Theorem 4.4.8] for part (b). Part (a) is easily derived from part (b). ∎

This theorem states that if the control objective is pole-assignment, then the corresponding controller design has the self-tuning property. The problem, that to the best of our knowledge is unsolved, is whether or not there are other controller designs that have the self-tuning property. This question has been studied in [3, 4] under some additional assumptions on the controller design. The assumptions there are introduced for the sake of finding an answer rather than from a systemtheoretic or even a mathematical motivation.
An alternative interpretation that we have for the first order case may be helpful in the analysis if it could be generalized to the higher-order case. We require that (35.6) implies $f(a, b) = f(a^0, b^0)$. This means that the curves defined by (35.6) and the equation $f(a, b) = f(a^0, b^0)$ should coincide. Since (a^0, b^0) belong to both curves, it follows that the curves should intersect tangentially in (a^0, b^0). This implies that f should satisfy the partial differential equation

$$f\frac{\partial f}{\partial a} + \frac{\partial f}{\partial b} = 0 \tag{35.13}$$

Since (a^0, b^0) is arbitrarily, every f that satisfies (35.13) has the self-tuning property.

5 References

[1] I.M.Y. Mareels and J.W. Polderman. *Adaptive Systems: an Introduction.* Birkhäuser, Boston, MA, 1996.

[2] J.W. Polderman. *Adaptive Control and Identification: Conflict or Conflux.* CWI Tract 67. Centre for Mathematics and Computer Science, Amsterdam, 1989.

[3] J.W. Polderman and C. Praagman. The closed-loop identification problem in indirect adaptive control. In *Proc. 28th IEEE Conf. Decision and Control*, 2120–2124, Tampa, Florida, USA, 1989.

[4] J.W. Polderman and C. Praagman. The closed-loop identification problem in indirect adaptive control. In K. Narendra, R. Ortega, and

P. Dorato, editors, *Advances in Adaptive Control*, 50–54. IEEE Press, 1989.

[5] J.W. Polderman and J.C. Willems. *Introduction to mathematical systems theory: a behavioral approach*, volume 26 of *Texts in Applied Mathematics*. Springer, New York NY, USA, 1997.

36

To estimate the L_2-gain of two dynamic systems

Anders Rantzer

Department of Automatic Control
Lund Institute of Technology
P.O. Box 118
S-221 00 LUND
SWEDEN
rantzer@control.lth.se

1 Description of the problem

1. Given numbers $\alpha, \beta, \delta > 0$ find the best possible upper bound on the L_2-induced gain from u to x that applies to all systems of the form

$$\dot{x}(t) = -\Phi(t)\Phi(t)'x(t) + u(t) \qquad x(0) = 0$$

where Φ satisfies $\alpha I \leq \int_s^{s+\delta} \Phi(t)\Phi(t)'dt \leq \beta I$ for all $s > 0$.

2. Given $k > 0$, determine the L_2-induced gain from u to $\ddot{x}$ in the system

$$\ddot{x} = k\ \mathrm{sat}(-x - \dot{x} + u) \qquad x(0) = 0,\ \dot{x}(0) = 0$$

In both problems, it is also interesting to construct inputs that give rise to large outputs. The problems can be generalized by letting the input enter through a term Bu and defining the output as linear combination Cx of the states.

2 Motivation

1. This type of time-varying systems appear in parameter estimation and adaptive control. For example, suppose that the time-varying coefficients $a(t)$ and $b(t)$ of the system $\dot{y}(t) = a(t)y(t) + b(t)w(t)$ are

to be estimated based on measurements of y and w. One approach is to update the estimates $\widehat{a}$ and $\widehat{b}$ according to the dynamics

$$\begin{bmatrix} d\widehat{a}/dt \\ d\widehat{b}/dt \end{bmatrix} = -\begin{bmatrix} y \\ w \end{bmatrix}(\widehat{a}y + \widehat{b}w - \dot{y})$$

Then the differential equation in problem 1 is satisfied by $\Phi = (y, w)$, $x = (a - \widehat{a}, b - \widehat{b})$ and $u = (da/dt, db/dt)$. The value of the L_2 gain gives quantitative information about the error in $\widehat{a}$ and $\widehat{b}$, given the amount of time-variation in a and b. It can also be used in combination with other integral quadratic constraints to analyse robustness of the esitimates, for example with respect to unmodeled dynamics or measurement errors in y and w.

2. Problems of this kind appear naturally in the study of control systems with saturations [6]. Recently, performance analysis of systems with rate limiters has been done based on a gain computation for a single integrator system in combination with well known integral quadratic constraints for saturations [4]. A solution of problem 2 would lead to similar applications for systems with acceleration bounds. Such bounds appear naturally for example in mechanical systems with magnitude bounds on control forces.

3 Available results

1. The system is known [1] to be exponentially stable for $u = 0$. Gain related arguments have been frequently used in the context of robust adaptive control. For an entry to this literature, see [2] and [3].

2. The system is asymptotically stable for $u = 0$, but the gain from u to x was proved in [5] to be infinite. This paper also gave a proof that the L_2-induced gain from u to x of the single integrator system

$$\dot{x} = \text{sat}(-x + u) \qquad x(0) = 0$$

is finite. The exact value $\sqrt{2}$ of this gain was computed in [4].

Acknowledgements

The author is grateful to A. Megretski for suggesting the second problem and to E. Sontag for providing valuable references.

4 References

[1] B.D.O. Anderson "Exponential Stability of Linear Equations Arising in Adaptive Identification", *IEEE Transactions on Automatic Control*, Feb. 1977, pp. 83-88.

[2] B.D.O. Anderson, R. R. Bitmead, C.R. Johnson, P.V. Kokotović, R.L. Kosut, I. Mareels, L. Praly and B. Riedle "Stability of Adaptive Systems", MIT Press, Cambridge, Massachusetts, 1986

[3] R. Ortega and T. Yu "Robustness of Adaptive Controllers: A Survey", *Automatica*, 25:5, pp. 651–678, 1989

[4] A. Rantzer and A. Megretski, "Analysis of Rate Limiters Using Integral Quadratic Constraints", Proceedings of IFAC Symposium on Nonlinear Control Systems Design (NOLCOS'98), Enschede, The Netherlands, June 1998

[5] W. Liu, Y. Chitour and E.D. Sontag, "On finite gain stabilizability of linear systems subject to input saturation, SIAM J. Control and Optimization, 34, pp. 1190-1219, 1996

[6] Sussmann, H.J., Sontag, E.D., Yang, Y., "A general result on the stabilization of linear systems using bounded controls, IEEE Transactions on Automatic Control, 39, pp. 2411-2425, 1994

37
Open problems in the area of pole placement

Joachim Rosenthal*[1] and Jan C. Willems**

*Department of Mathematics
University of Notre Dame
Notre Dame, IN 46556
USA
Rosenthal.1@nd.edu

**Mathematics Institute
University of Groningen
P.O. Box 800
9700 AV Groningen
THE NETHERLANDS
willems@math.rug.nl

1 Introduction

The static and the dynamic output pole placement problem belong to the prominent design problems of modern control theory and we refer to the survey articles [4, 10, 19, 21] where also more references to the literature are provided. Various facets of the pole placement problem attracted many researchers over the years. It was been recognized right at the beginning of the problem that the output pole placement problem is nonlinear in nature and a simple solution based on techniques from linear algebra cannot be expected. In recent years significant progress has been achieved. This progress is due in a major part to a better understanding of the system theoretic ingredients and its relation to algebraic geometry. Helpful in this regard is the behavioral approach which comes in its formulation closest to the algebraic geometric nature of the pole placement problem.
Classically, the static output feedback problem is formulated in the following way: Let A, B, C be real matrices of size $n \times n$, $n \times m$, and $p \times n$,

[1]Supported in part by NSF grant DMS-96-10389.

respectively. These 3 matrices describe the linear system

$$\frac{d}{dt}x = Ax + Bu,\ y = Cx. \tag{37.1}$$

Let $\phi \in \mathbb{R}[\xi]$ be a monic polynomial of degree n. Then the static pole placement problem asks for conditions which guarantee the existence of a $m \times p$ matrix K such that

$$\det(\xi I - A - BKC) = \phi(\xi).$$

After elimination of the latent variable x, equation (37.1) describes a linear differential system. This elimination can be done as follows in the controllable case. If $D(\xi)^{-1}N(\xi)$ is a left coprime factorization of the transfer function $C(\xi I - A)^{-1}B$ then the manifest behavior is described by

$$\left(D\left(\frac{d}{dt}\right) \quad -N\left(\frac{d}{dt}\right)\right)\begin{pmatrix} y \\ u \end{pmatrix} = 0. \tag{37.2}$$

It is easy to verify that the design of a static compensator is equivalent to the construction of matrices K_1, K_2 such that

$$\det\begin{bmatrix} D(\xi) & -N(\xi) \\ K_1 & K_2 \end{bmatrix} = \phi(\xi). \tag{37.3}$$

In the next section we summarize some of the major known pole placement results. The essence of the problems and their solutions are most transparent in the behavioral language. For the connection between the behavioral point of view and the classical state space formulation we refer to [14, 25]. The following development of the theory follows closely the description given in [18].

2 A summary of known pole placement results

Recall from [14] that a *dynamical system* Σ is a triple $\Sigma = (T, W, \mathcal{B})$, where $T \subset \mathbb{R}$ is the time axis, W is the signal space and $\mathcal{B} \subset W^T$ is called the behavior. We restrict attention sequel to dynamical systems Σ with time axis $T = \mathbb{R}$, signal space $W = \mathbb{R}^{m+p}$, and behavior $\mathcal{B} \subset C^\infty(\mathbb{R}, \mathbb{R}^{m+p})$ specified by a so called *kernel representation*, i.e. there exist a polynomial matrix P such that

$$\mathcal{B} = \{w \in C^\infty(\mathbb{R}, \mathbb{R}^{m+p}) \mid P\left(\frac{d}{dt}\right)w = 0\}. \tag{37.4}$$

Systems in this class are called *linear differential systems.* Such a behavior $\mathcal{B}$ admits many kernel representations (37.4) (see [14]). Two important

invariants are the *rank* $r(\Sigma)$ and the *McMillan degree* $n(\Sigma)$, which are defined as follows: $r(\Sigma)$ is equal to the rank of a polynomial matrix P which describes the behavior $\mathcal{B}$ by a representation of the form (37.4). We will call a representation P *(row) minimal* if the matrix P has full row rank. If the polynomial matrix P is row minimal then we define the McMillan degree $n(\Sigma)$ as the maximal degree of the full size minors in one (and therefore any) minimal representation.
A system $\Sigma = (\mathbb{R}, \mathbb{R}^{m+p}, \mathcal{B})$ is *autonomous* if the rank $r(\Sigma) = m + p$. If the $(m+p) \times (m+p)$ matrix P describes an autonomous behavior then the nonzero polynomial $\det P$ is called the *characteristic polynomial* of Σ and will be abbreviated with χ_Σ; χ_Σ is a projective invariant of the autonomous behavior, i.e. if

$$\det P(\xi) = a_0 + a_1\xi + \cdots + a_n\xi^n$$

then $(a_0, \ldots, a_n)$ defines a unique one dimensional subspace of polynomials, i.e. a point in the projective space $\mathbb{P}^n$. This point then only depends on the autonomous system Σ and not on the particular representation. The roots of $\det P$ are by definition the *poles* of Σ.
Control in the behavioral theory is defined in the following way: Assume that $\Sigma_1 = (\mathbb{R}, \mathbb{R}^{m+p}, \mathcal{B}_1)$ and $\Sigma_2 = (\mathbb{R}, \mathbb{R}^{m+p}, \mathcal{B}_2)$ are two linear differential systems. Then the *interconnected system* $\Sigma_1 \wedge \Sigma_2$ is defined as:

$$\Sigma_1 \wedge \Sigma_2 := (\mathbb{R}, \mathbb{R}^{m+p}, \mathcal{B}_1 \cap \mathcal{B}_2).$$

We say $\Sigma_1 \wedge \Sigma_2$ is a *regular* or *independent* interconnection (see [25]) if the ranks "add up", i.e. if

$$r(\Sigma_1 \wedge \Sigma_2) = r(\Sigma_1) + r(\Sigma_2)$$

and we speak of a *singular* or *dependent* interconnection otherwise. As it is immediate from the definition the interconnected system $\Sigma_1 \wedge \Sigma_2$ is represented by

$$\begin{pmatrix} P_1\left(\frac{d}{dt}\right) \\ P_2\left(\frac{d}{dt}\right) \end{pmatrix} w = 0,$$

where P_1, P_2 are polynomial matrices representing Σ_1 respectively Σ_2. If P_1, P_2 are in addition row minimal representations then one verifies that $\Sigma_1 \wedge \Sigma_2$ is a regular interconnection if and only if $\binom{P_1}{P_2}$ is row minimal.
The fundamental question, which is in fact a generalized pole placement problem and which implies many of the "traditional" pole placement questions, is now as follows:

Problem 4 Let m, p, n, q be fixed positive integers. Under what condition is it true that for a generic set of linear differential systems $\Sigma_1 = (\mathbb{R}, \mathbb{R}^{m+p}, \mathcal{B}_1)$ having rank $r(\Sigma_1) = p$ and McMillan degree $n(\Sigma_1) = n$ the following holds: For every polynomial $\phi \in \mathbb{R}[\xi]$ of degree $n + q$ there exists a system $\Sigma_2 = (\mathbb{R}, \mathbb{R}^{m+p}, \mathcal{B}_2)$ having rank $r(\Sigma_2) = m$ and McMillan degree $n(\Sigma_2) = q$ such that $\Sigma_1 \wedge \Sigma_2$ forms a regular interconnection having characteristic polynomial $\chi_{\Sigma_1 \wedge \Sigma_2} = \phi$.

Problem 4 is the behavioral formulation of the dynamic pole placement problem. Closely related is the following formulation, which can be done for an arbitrary base field $\mathbb{F}$:

Problem 5 Let $\mathbb{F}$ be a field and let m, p, n, q be fixed positive integers. Let P_1 be a $p \times (m+p)$ matrix with entries in $\mathbb{F}[\xi]$ whose $p \times p$ minors have degree at most n. Let $\phi(\xi) \in \mathbb{F}[\xi]$ be a polynomial of degree $n + q$. Under what conditions does there exist a $m \times (m + p)$ matrix P_2 with entries in $\mathbb{F}[\xi]$ such that the $m \times m$ minors of P_2 have degree at most q and

$$\det \begin{pmatrix} P_1 \\ P_2 \end{pmatrix} = \phi?$$

If Problem 5 has a positive answer for a 'generic set' of $p \times (m + p)$ matrices of degree n then one says that the generic rank p system of McMillan degree n is *arbitrary pole assignable* in the class of feedback compensators of McMillan degree q. If $q = 0$ one says that the generic rank p system of McMillan degree n is arbitrary pole assignable by static feedback compensators. Of course it is a major difficulty to make the notion of genericity precise in this formulation and we refer to [18] for details on this question.

2.1 *Static pole placement results*

The major results in the area of static pole placement are as follows:

Theorem 6 (Brockett and Byrnes [3]) *If the base field $\mathbb{F}$ is algebraically closed and if $mp \geq n$ then the generic rank p system of McMillan degree n is arbitrary pole assignable by static feedback compensators. Moreover if $mp = n$ then (counting multiplicity) the number of non-equivalent feedback compensators assigning a particular closed loop characteristic polynomial is independent of the closed loop polynomial $\phi(\xi) \in \mathbb{F}[\xi]$ and is equal to*

$$d(m,p) = \frac{1!2! \cdots (p-1)!(mp)!}{m!(m+1)! \cdots (m+p-1)!} \tag{37.5}$$

Since $mp \geq n$ is easily seen to be a necessary condition, Theorem 6 gives the best possible bound when the base field $\mathbb{F}$ is algebraically closed.
The number $d(m,p)$ is the degree of the Grassmann variety, which was computed in the last century by Schubert [20]. Although Theorem 6 assumes that the base field $\mathbb{F}$ is algebraically closed it does also provide some results for real pole assignment:

Corollary 7 *If $\mathbb{F} = \mathbb{R}$, $mp = n$, and $d(m,p)$ is odd, then the generic rank p system of McMillan degree n is arbitrary pole assignable by static real feedback compensators.*

Berstein determined when $d(m,p)$ is odd.

Proposition 8 (Berstein [1]) *The number $d(m,p)$ is odd if and only if $\min(m,p) = 1$ or $\min(m,p) = 2$ and $\max(m,p) = 2^t - 1$, where t is a positive integer.*

When $d(m,p)$ is even, the best known sufficiency result over the reals is due to Wang:

Theorem 9 (Wang [22]) *If $\mathbb{F} = \mathbb{R}$ and $mp > n$, then the generic rank p system of McMillan degree n is arbitrary pole assignable by static feedback compensators.*

In the last 3 years several elementary proofs of Wang's theorem were given [9, 15, 16, 26].
When the Schubert number $d(m,p)$ is even, there is a difference of one degree of freedom between the sufficiency condition of Wang ($mp > n$) and the general necessary condition ($mp \geq n$). The following lemma states that in general $mp \geq n$ is not a sufficient condition for generic pole placement with real static compensators:

Lemma 10 (Willems and Hesselink [27]) *If $\mathbb{F} = \mathbb{R}$ and if $m = p = 2$ and $n = 4$ then the static pole placement problem is not generically solvable over the reals $\mathbb{R}$.*

2.2 Dynamic pole placement results

Also for the dynamic problem there are some general necessary conditions. A simple dimension argument first carried out in [27] reveals that

$$q(m + p - 1) + mp \geq n \tag{37.6}$$

is a necessary condition for the generic rank p system of McMillan degree n to be arbitrary pole assignable in the class of feedback compensators of McMillan degree q. In [13, 14] the sufficiency result of Theorem 6 was extended:

Theorem 11 (Rosenthal-Ravi-Wang [13, 14]) *If the base field $\mathbb{F}$ is algebraically closed and if $q(m + p - 1) + mp \geq n$ then the generic rank p system of McMillan degree n is arbitrary pole assignable in the class of feedback compensators of degree at most q. Moreover if there is equality in (37.6) then the number of non-equivalent feedback compensators assigning a particular closed loop characteristic polynomial is independent of the closed loop polynomial $\phi(\xi) \in \mathbb{F}[\xi]$ and equals (counting multiplicity) to*

$$d(m,p,q) =$$

$$(-1)^{q(m+1)}(mp+q(m+p))! \sum_{n_1+\cdots+n_m=q} \frac{\prod_{k<j}(j-k+(n_j-n_k)(m+p))}{\prod_{j=1}^{m}(p+j+n_j(m+p)-1)!} \tag{37.7}$$

Note that the number $d(m,p,0)$ is equal to $d(m,p)$ as introduced in (37.5). As in the case of the static pole placement problem condition (37.6) is also sufficient if $d(m,p,q)$ is odd. If $d(m,p,q)$ is even the strongest known sufficiency result over the reals is:

Theorem 12 (Rosenthal-Wang [18]) *Let* $\mathbb{F} = \mathbb{R}$ *and assume that*

$$q(m+p-1)+mp-\min(r_m(p-1), r_p(m-1)) > n \tag{37.8}$$

where $r_m = q - m[q/m]$ *and* $r_p = q - p[q/p]$ *are the remainders of* q *divided by* m *and* p*, respectively. Then the generic rank* p *system of McMillan degree* n *is arbitrary pole assignable in the class of real feedback compensators of degree at most* q*.*

3 A list of open problems

3.1 Necessary and sufficient conditions for generic pole assignment over the reals

It has been conjectured by S.-W. Kim that $m = p = 2$, $n = 4$ is the only case where $mp = n$ is not a sufficient condition for generic pole assignment with static compensators over the reals. This conjecture was disproved by Rosenthal and Sottile in [17] by exhibiting a concrete counterexample in the situation where $m = 2$ and $p = 4$. This lead to the following conjecture:

Conjecture 13 (Rosenthal-Sottile [17]) *If* $d(m,p)$ *is even and* $n = mp$*, then the static pole placement problem is not generically solvable over the reals* $\mathbb{R}$*.*

This conjecture is open. For the general dynamic problem much less is known and the gap between the general necessary condition (37.6) and the best known sufficiency condition (37.8) is in general quite wide and it can be as large as $(m-1)(p-1)$. It is a challenging task to further narrow this gap, and obtain a sharp bound as Theorem 6. In addition, it is of interest to have exact algebraic criteria which allow to determine the minimal order of a real compensator needed, over the reals. For the static pole placement problem such algebraic conditions have been recently derived in [17].

3.2 *Numerical algorithms for pole placement*

The sufficiency conditions in Theorems 6,9,11 and 12 are mainly theoretical in nature and there are no good numerical algorithms available in many situations where we know the existence of a feedback compensator. Theorem 9 and 12 were derived by a technique called linearization around a 'dependent compensator'. In principle this technique can be used to actually compute feedback compensators (see [18, 23] for more details). Geometrically dependent compensators form the *base locus* (see [14]) of the associated pole placement map and the numerical paradigm should probably be to construct solutions which are as far as possible away from the base locus. In principle it is also possible to tackle any pole placement problem directly through the defining polynomial equations and applying methods like e.g. Gröbner basis computations. Such an approach was taken in [17] but it is clear that computations will be limited to small dimensions.
In summary we consider it a challenge to derive stable numerical algorithms to solve pole placement problems.

3.3 *The invariant polynomial assignment problem*

The characteristic polynomial forms in general not the only invariant of an autonomous behavior. A finer set of invariants are the invariant factors. Two autonomous behaviors are isomorphic if and only if the polynomial matrices in a row minimal representation have the same invariant factors. This follows readily from the Smith forms of these polynomial matrices. It is therefore reasonable to ask for necessary and sufficient conditions which guarantee the assignment of a set of invariant factors for a system $\Sigma_1 = (\mathbb{R}, \mathbb{R}^{m+p}, \mathcal{B}_1)$ having rank $r(\Sigma_1) = p$. We refer to [25] for details.

3.4 *Feedback stabilization versus pole placement*

It has been shown in [5, 11] that if a system is not generically pole assignable then there exist an open set of systems (open in the Euclidean topology) which cannot be stabilized. Based on those results one might think that the study of pole assignability does cover the question of stabilizability. This is only partially true. We believe that it is worthwhile to have results which allows one to describe the set of systems of a fixed McMillan degree that cannot be stabilized by regular feedback interconnections of compensators of degree at most q.

3.5 *The problem of simultaneous pole placement and simultaneous stabilization*

Simultaneous pole placement tries to answer the following question: Given a set of plants $\Sigma_i = (\mathbb{R}, \mathbb{R}^{m+p}, \mathcal{B}_i)$, $i = 1, \ldots, r$ degree $n(\Sigma_i) = n_i$ and

having constant rank $r(\Sigma_1) = p$. Let ϕ_i be a set of polynomials of degree $n_i + q$. We are seeking a feedback compensator $\Sigma = (\mathbb{R}, \mathbb{R}^{m+p}, \mathcal{B})$ of degree q and rank $r(\Sigma) = m$ such that all feedback interconnections $\Sigma_i \wedge \Sigma$ are regular and the characteristic polynomial $\chi_{\Sigma_i \wedge \Sigma} = \phi_i$.
In the work of Ghosh [6] some sufficiency results were derived. Those results are however far from being optimal. We would like to note that the question is an algebraic problem and this question is fully decidable. This is different from the simultaneous stabilization problem studied by Blondel [2].

3.6 The problem of decentralized pole placement

Let P describe a linear differential system of rank p and McMillan degree n. The decentralized pole placement problem asks for the construction of a set of compensators $Q_1, \ldots, Q_r$ such that a desired closed loop characteristic polynomial of the form

$$\phi(\xi) = \det \begin{bmatrix} & P(\xi) & & \\ \hline Q_1(\xi) & 0 & \ldots & 0 \\ 0 & Q_2(\xi) & \ldots & 0 \\ \vdots & \vdots & \ddots & 0 \\ 0 & 0 & \ldots & Q_r(\xi) \end{bmatrix} \tag{37.9}$$

can be achieved. In [12] necessary and sufficient conditions over the complex numbers were derived. There are however no strong results for real pole assignment known.

3.7 General matrix extension problems

It has been explained in [19] that many pole placement problems can be expressed as matrix extension problems. It would be desirable to have strong necessary and strong sufficient conditions available which cover a wide range of matrix extension problems. A first general result in this direction has been derived in [15]. For example, consider an $n \times n$ matrix of the form

$$\begin{bmatrix} * & \cdots & * & * & \cdots & * & * & \cdots & * \\ \vdots & \ddots & \vdots & \vdots & \ddots & \vdots & \vdots & \ddots & \vdots \\ * & \cdots & * & * & \cdots & * & * & \cdots & * \\ * & \cdots & * & ? & \cdots & ? & * & \cdots & * \\ \vdots & \ddots & \vdots & \vdots & \ddots & \vdots & \vdots & \ddots & \vdots \\ * & \cdots & * & ? & \cdots & ? & * & \cdots & * \\ * & \cdots & * & * & \cdots & * & * & \cdots & * \\ \vdots & \ddots & \vdots & \vdots & \ddots & \vdots & \vdots & \ddots & \vdots \\ * & \cdots & * & * & \cdots & * & * & \cdots & * \end{bmatrix} \tag{37.10}$$

where the $*$'s signify known elements, and the ?'s are elements that can be chosen freely. Is it possible to choose, for any monic n–th order $\phi \in \mathbb{F}[\xi]$, and for generic numbers $*$'s, the numbers ?'s such that the matrix has characteristic polynomial ϕ, provided the number of ?'s exceeds n? If $\mathbb{F} = \mathbb{C}$ this has been answered in [15]. Over the reals the question is open. Note that the problem is almost, but not quite, a consequence of Theorem 6.

4 References

[1] I. Berstein. On the Lusternik-Šnirel'mann category of real Grassmannians. *Proc. Camb. Phil. Soc.*, 79:129–239, 1976.

[2] V. Blondel. *Simultaneous Stabilization of Linear Systems*, volume 191 of *Lecture Notes in Control and Information Sciences.* Springer-Verlag London Ltd., London, 1994.

[3] R. W. Brockett and C. I. Byrnes. Multivariable Nyquist criteria, root loci and pole placement: A geometric viewpoint. *IEEE Trans. Automat. Control*, AC-26:271–284, 1981.

[4] C. I. Byrnes. Pole assignment by output feedback. In H. Nijmeijer and J. M. Schumacher, editors, *Three Decades of Mathematical System Theory*, Lecture Notes in Control and Information Sciences, volume 135, pages 31–78. Springer Verlag, 1989.

[5] C. I. Byrnes and B. D. O. Anderson. Output feedback and generic stabilizability. *SIAM J. Control Optim.*, 22(3):362–380, 1984.

[6] B. K. Ghosh. An approach to simultaneous system design. part II: Nonswitching gain & dynamic feedback compensation by algebraic geometric methods. *SIAM J. Control Optim.*, 26(4):919–963, 1988.

[7] W. Helton, J. Rosenthal, and X. Wang. Matrix extensions and eigenvalue completions, the generic case. *Trans. Amer. Math. Soc.*, 349(8):3401–3408, 1997.

[8] N. Karcanias and C. Giannakopoulos. Grassmann invariants, almost zeros and the determinantal zero, pole assignment problems of linear multivariable systems. *Internat. J. Control*, 40(4):673–698, 1984.

[9] J. Leventides and N. Karcanias. Global asymptotic linearization of the pole placement map: A closed form solution for the constant output feedback problem. *Automatica*, 31(9):1303–1309, 1995.

[10] N. Munro. Pole assignment: A review of methods. In M. G. Singh, editor, *Systems and Control Encyclopedia*, pages 3710–3717. Pergamon Press, 1990.

[11] M. S., J. Rosenthal, and X. Wang. On generic stabilizability and pole assignability. *Systems & Control Letters*, 23(2):79–84, 1994.

[12] M. S. Ravi, J. Rosenthal, and X. Wang. On decentralized dynamic pole placement and feedback stabilization. *IEEE Trans. Automat. Contr.*, 40(9):1603–1614, 1995.

[13] M. S. Ravi, J. Rosenthal, and X. Wang. Dynamic pole assignment and Schubert calculus. *SIAM J. Control Optim.*, 34(3):813–832, 1996.

[14] J. Rosenthal. On dynamic feedback compensation and compactification of systems. *SIAM J. Control Optim.*, 32(1):279–296, 1994.

[15] J. Rosenthal, J. M. Schumacher, X. Wang, and J. C. Willems. Generic eigenvalue assignment for generalized linear first order systems using memoryless real output feedback. In *Proc. of the 34th IEEE Conference on Decision and Control*, pages 492–497, New Orleans, Louisiana, 1995.

[16] J. Rosenthal, J. M. Schumacher, and J. C. Willems. Generic eigenvalue assignment by memoryless real output feedback. *Systems & Control Letters*, 26:253–260, 1995.

[17] J. Rosenthal and F. Sottile. Some remarks on real and complex output feedback. *Systems & Control Letters*, 33(2):73–80, 1998.

[18] J. Rosenthal and X. Wang. Output feedback pole placement with dynamic compensators. *IEEE Trans. Automat. Contr.*, 41(6):830–843, 1996.

[19] J. Rosenthal and X. Wang. Inverse eigenvalue problems for multivariable linear systems. In C. I. Byrnes, B. N. Datta, D. Gilliam, and C. F. Martin, editors, *Systems and Control in the Twenty-First Century*, pages 289–311. Birkäuser, 1997.

[20] H. Schubert. Beziehungen zwischen den linearen Räumen auferlegbaren charakteristischen Bedingungen. *Math. Ann.*, 38:598–602, 1891.

[21] V. L. Syrmos, C. T. Abdallah, P. Dorato, and K. Grigoriadis. Static output feedback—a survey. *Automatica*, 33(2):125–137, 1997.

[22] X. Wang. Pole placement by static output feedback. *Math. Systems, Estimation, and Control*, 2(2):205–218, 1992.

[23] X. Wang. Grassmannian, central projection and output feedback pole assignment of linear systems. *IEEE Trans. Automat. Contr.*, 41(6):786–794, 1996.

[24] J. C. Willems. Paradigms and puzzles in the theory of dynamical systems. *IEEE Trans. Automat. Control*, AC-36(3):259–294, 1991.

[25] J. C. Willems. On interconnections, control, and feedback. *IEEE Trans. Automat. Control*, 42(3):326–339, 1997.

[26] J. C. Willems. Generic eigenvalue assignability by real memoryless output feedback made simple. A. Paulraj, V. Roychowdhury, and C.D. Schaper, editors, In *Communications, Computation, Control and Signal Processing*, Kluwer, pages 343-354, 1997.

[27] J. C. Willems and W. H. Hesselink. Generic properties of the pole placement problem. In *Proc. of the 7th IFAC Congress*, pages 1725–1729, 1978.

38

An optimal control theory for systems defined over finite rings

Joachim Rosenthal[1]

Department of Mathematics
University of Notre Dame
Notre Dame, IN 46556
USA
Rosenthal.1@nd.edu

1 Introduction

A fundamental problem of coding theory is the efficient decoding of certain classes of convolutional codes. It would be highly desirable to develop efficient algorithms which are capable of decoding classes of convolutional codes with particular algebraic properties. In this article we do formulate the problem in terms of linear systems theory and we show that the posed question is connected to some classical problems of systems theory.
A convolutional code can be viewed as a discrete time linear system defined over a finite field $\mathbb{F}$ and we will say more about it in Section 2. Sometimes it is too restrictive to work over a finite field $\mathbb{F}$ and because of this several authors did recently consider codes over a finite ring R (such as the ring $\mathbb{Z}_q$ consisting of the integers modulo q e.g.) or even codes over arbitrary finite groups (see e.g. [3]).
Convolutional codes are widely used in the transmission of data over noisy channels. In conjunction with data compression and modulation schemes they are nowadays integral part of many communication devices. As an example we want to mention the transmission of pictures and other data from deep space, where NASA has used convolutional codes in a most successful way.
In the literature one can find several decoding algorithms. Probably the most widely implemented algorithm is the Viterbi decoding algorithm and we refer to the textbooks [2, 13] for details. Under some natural assumption

[1]Supported in part by NSF grant DMS-96-10389.

on the statistics of the error pattern this algorithm is capable of decoding a received message in a 'maximum likelihood' fashion. The disadvantage of this algorithm lies in the fact that practically the algorithm is too complex for convolutional codes whose McMillan degree is more than 20. On the side of the Viterbi algorithm there exist several 'suboptimal algorithms'. These algorithms do in general not compute the code word in a maximum likelihood fashion. We refer again to the textbooks [2, 6, 13].

The reader might wonder why we pose the decoding problem and why we believe that progress can be done in this area. The reasons are as follows. First note that convolutional codes naturally generalize block codes. Indeed we can view a block code as a convolutional code of McMillan degree zero. For block codes there exist a wealth of algebraic decoding algorithms which take advantage of the algebraic properties of the block code. In contrast to the situation of block codes convolutional codes of nonzero McMillan degree are typically found by computer searches and the existing algorithms do not take advantage of any algebraic structure. Actually most books in coding theory treat convolutional codes in a mainly graph theoretical way and systems theoretic properties of the code are only remarked on the side. It is the author's believe that it should be possible to algebraically construct convolutional codes (linear systems) which come in conjunction with some powerful decoding algorithm. Such an algorithm most likely will employ systems theoretic properties of the underlying code. A first attempt to carry through such a program was reported in [10]. We also see the possibility that existing algorithms in the area of filtering [1] and modeling [7, 8] might lead to improvements in the area of decoding.

The paper is structured as follows: In the next section we introduce the class of convolutional codes defined over a finite ring R. In Section 3 we explain the decoding problem in the situation where the data has been transmitted over the so called q-ary symmetric channel. Finally we explain in Section 4 the decoding problem if data has been transmitted over a Gaussian channel.

We did make an attempt that the paper is self contained. Because of space limitations we did present the problem as a systems theoretic problem. The reader interested in issues of coding theory is referred to the literature. A standard reference on convolutional codes is the textbook by Lin and Costello [2]. The algebraic structure of convolutional codes in the way it is treated in the coding literature is probably best described in the monograph of Piret [6]. One of the most comprehensive reference on linear block codes is the book by MacWilliams and Sloane [4]. The connection of convolutional codes to linear systems theory was first recognized by Massey and Sain [5]. More details on this connection and the way we present the problem are given in the recent papers [9, 11, 12] and the dissertation of York [16].

2 Convolutional codes defined over a Galois ring R

Let R be a finite ring. R is sometimes referred to as a *Galois ring* since this class of rings naturally generalizes the class of Galois fields.
It is the goal of coding theory to transmit data over some noisy channel. For this assume that a vector $v_t \in R^n$ is transmitted at time $t = 0, 1, 2, \ldots$. In this way we arrive at a time series

$$v = \{v_0, v_1, v_2, \ldots\} \in (R^n)^{\mathbb{Z}_+}. \tag{38.1}$$

In order to allow the possibility of error correction it will be necessary to restrict the set of all possible trajectories in $(R^n)^{\mathbb{Z}_+}$ to some subset $\mathcal{C}$ and add in this way some redundancy. A natural way to do such a restriction is as follows:
A subset $\mathcal{C} \subset (R^n)^{\mathbb{Z}_+}$ is called *right shift invariant* if

$$\{v_0, v_1, v_2, \ldots\} \in \mathcal{C} \Longrightarrow \{0, v_0, v_1, v_2, \ldots\} \in \mathcal{C}.$$

The property of right shift invariance allows a time delay in the transmission of the data without confusing the receiver.
Set theoretically $(R^n)^{\mathbb{Z}_+}$ is isomorphic to the direct product $\prod_{i=0}^{\infty} R^n$. In this way $(R^n)^{\mathbb{Z}_+}$ has a natural R-module structure and we define:

Definition 14 A R-linear and right shift invariant subset $\mathcal{C} \subset (R^n)^{\mathbb{Z}_+}$ is called a convolutional code.

The following two examples illustrate two important cases of convolutional codes.

Example 15 Assume $M \subset R^n$ is a R submodule of R^n. If $\mathcal{C}$ is of the form

$$\mathcal{C} = \prod_{i=0}^{\infty} M \subset \prod_{i=0}^{\infty} R^n \cong (R^n)^{\mathbb{Z}_+}$$

then we call $\mathcal{C}$ a linear block code. Alternatively $\mathcal{C}$ consists of all sequences $v = \{v_0, v_1, v_2, \ldots\} \in (R^n)^{\mathbb{Z}_+}$ having the property that $v_t \in M, t = 0, 1, 2 \ldots$. One disadvantage of block codes lies in the fact that so called *burst errors*, these are errors which affect a whole block, are in general badly protected unless the block size is very large. Despite this disadvantage block codes are widely implemented and there are many known techniques of constructing and decoding block codes even if the block length n is very large (see e.g. [4]).

Example 16 The set of code words are often generated by particular input-output systems. For this consider matrices A, B, C, D with entries in R and consider the discrete time system defined over R:

$$x_{t+1} = Ax_t + Bu_t, \quad y_t = Cx_t + Du_t, \quad x_0 = 0. \tag{38.2}$$

Equation (38.2) forms the state space realization of a 'systematic encoder'. (Compare with [2, 9]). One verifies that the collection of all possible trajectories

$$v_t := \begin{pmatrix} u_t \\ y_t \end{pmatrix}, \quad t = 0, 1, 2, \ldots$$

defines a convolutional code $\mathcal{C}$. If the matrices A, B, C, D have size $\delta \times \delta$, $\delta \times k$, $(n-k) \times \delta$ and $(n-k) \times k$ respectively one says that $\mathcal{C}$ has complexity (=McMillan degree) δ and transmission rate k/n. Convolutional codes having the form (38.2) are very convenient since in the encoding process u_t can be chosen freely whereas y_t describes the added redundancy. This explains also the word transmission rate since for every k symbols n symbols have to be transmitted.
If the complexity $\delta = 0$ there is the same linear constraint $y_t = Du_t$ at each time instance t and we deal again with a linear block code.

The reader who is familiar with the behavioral literature [14] will observe the close connection to the presented approach. We would like however to stress that our definition of convolutional code does not quite coincide with the notion of linear behavior of Willems [14]. Indeed we have not imposed (and we follow here [3, 11]) that the code has to be complete, a basic requirement for a linear behavior.
Instead of imposing completeness one might want to impose that a code sequence $v = \{v_t\}_{t \in \mathbb{Z}_+}$ has finite support, i.e. v_t is zero with the exception of finitely many time instances. This approach has been taken in [11, 12, 16] and it is based on the reasoning that every data transmission will end at some time. By requiring that a convolutional code has finite support we achieve a duality between convolutional codes on one side and linear behaviors on the other side and we refer to [11] for details. In particular it will still be possible to employ known systems theoretic descriptions for convolutional codes.

3 The problem of decoding convolutional codes on the symmetric channel

The optimal way of decoding a convolutional code depends on the error statistics of the transmission channel. In this section we explain the decoding problem if the transmission channel consists of the so called *q-ary symmetric channel* which we will define in a moment:
Assume the ring R consists of the q symbols $r_1, \ldots, r_q$. The q-ary symmetric channel assumes that during the transmission process every element $r_j, j = 1, \ldots, q$ might change into some element different of r_j with some fixed probability p. In this way the receiver will obtain a time series $\hat{v} = \{\hat{v}_t\}_{t \in \mathbb{Z}_+} \subset (R^n)^{\mathbb{Z}_+}$ and the decoding task is to find the time series

$v \in C$ which comes 'closest' to the received time series $\hat{v}$. In order to specify what close means in our context we will have to introduce the notion of *Hamming metric*:
If $w \in R^n$ is any vector one defines its Hamming weight as the number of nonzero components of the n-vector w. We will denote the Hamming weight of w by $\mathrm{Ham}(w)$. If $w, \hat{w} \in R^n$ are any two vectors one defines their Hamming distance through the formula $\mathrm{dist}(w, \hat{w}) := \mathrm{Ham}(w - \hat{w})$. One immediately verifies that 'dist' satisfies all axioms of a metric on the (finite) set R^n.
Assume that a certain code word $v = \{v_t\}_{t\in\mathbb{Z}_+}$ was sent and that the message word $\hat{v} = \{\hat{v}_t\}_{t\in\mathbb{Z}_+}$ has been received. The decoding problem then asks for the minimization of the error

$$\text{error} := \min_{v\in C} \sum_{t\in\mathbb{Z}_+} \mathrm{dist}(v_t, \hat{v}_t). \tag{38.3}$$

In the concrete setting of Example 16 the decoding problem asks for the minimization of the error

$$\text{error} = \min \left(\sum_{t=0}^{\infty} (\mathrm{dist}(u_t, \hat{u}_t) + \mathrm{dist}(y_t, \hat{y}_t)) \right), \tag{38.4}$$

where we denote as before with $\{v_t\}_{t\geq 0} = \left\{ \binom{u_t}{y_t} \right\}_{t\geq 0}$ a particular code word and with $\{\hat{v}_t\}_{t\geq 0} = \left\{ \binom{\hat{u}_t}{\hat{y}_t} \right\}_{t\geq 0}$ a received message word.
Theoretically a correct decoding can always be achieved as long as the error magnitude is at most 1/2 of the so called *free distance*. The free distance of a code measures the smallest distance between any two different code words and it is formally defined as:

$$d_{\text{free}} := \min_{\substack{u,v\in C \\ u\neq v}} \sum_{t\in\mathbb{Z}_+} \mathrm{dist}(u_t, v_t). \tag{38.5}$$

Note that the decoding problem is essentially a discrete 'tracking problem' where the received message word $\hat{v} = \{\hat{v}_t\}_{t\geq 0}$ has to be tracked by the 'nearest valid code word'. If no transmission error did occur then $\hat{v} = \{\hat{v}_t\}_{t\geq 0}$ is a valid trajectory and the error value in (38.4) can be made zero. It is also possible to view the decoding problem as a 'discrete filtering problem' where the error sequence $e_t := \hat{v}_t - v_t$ has to be filtered out from the received sequence $\hat{v}_t$. In the literature about digital filters (see e.g. [1]) one finds sometimes the appropriate term 'deconvolution'. The problem as we pose it here is formally also closely connected to a global least square modeling problem as it has been recently studied by Roorda [7] and Roorda and Heij [8].
Unfortunately it is not easy to connect either to above literature since the underlying metric is not the Euclidean metric but rather the Hamming metric. As mentioned in the introduction the predominant algorithm applied in

the coding area is the Viterbi decoding algorithm. This algorithm applies the principal of dynamic programming to above situation. The Viterbi algorithm is always applicable but it becomes computationally infeasible as soon as the complexity of the encoder (38.2) is more than a fairly small number like 20. Indeed the number of possible states of an encoder with complexity δ is q^δ, where q is the cardinality of the ring R. The Viterbi algorithm requires a search in a graph which has more than q^δ vertices and this is in terms of complexity not feasible if the complexity and the cardinality of R are too large.

4 Decoding on the Gaussian channel

In many transmission situations such as transmissions over telephone lines and transmission in deep space the signal alphabet is mapped into points of the complex plane $\mathbb{C}$. If $re^{i\theta} \in \mathbb{C}$ is a particular point in the complex plane the transmission is done by assigning to the signal a phase angle of θ and an amplitude of r. The mapping of the signal alphabet into the complex plane $\mathbb{C}$ (or even into some Cartesian product $\mathbb{C}^i$ of $\mathbb{C}$) is called a *modulation.* In the sequel we explain the most basic of these ideas and we refer the interested reader to the textbooks [13, 15] for further reading.

In practice there are two widely implemented modulation schemes. The first is called *q-ary phase shift keying* abbreviated by q-PSK. In this modulation scheme the amplitude of each signal is the same and the modulation is done by assigning to each letter of the alphabet R some phase angle.

The second method is called *q-ary amplitude modulation* usually abbreviated by q-AM. In this scheme the phase angle is left constant. The following picture depicts a typical phase shift modulation scheme and a typical amplitude modulation scheme.

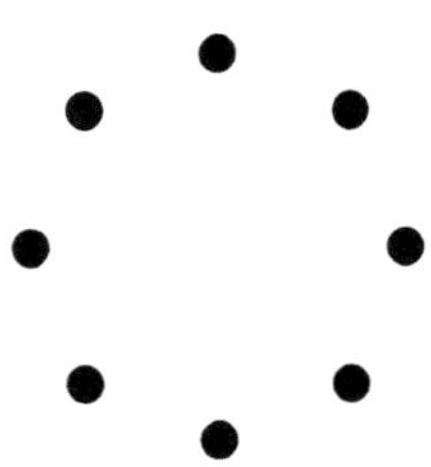

8-PSK Constellation

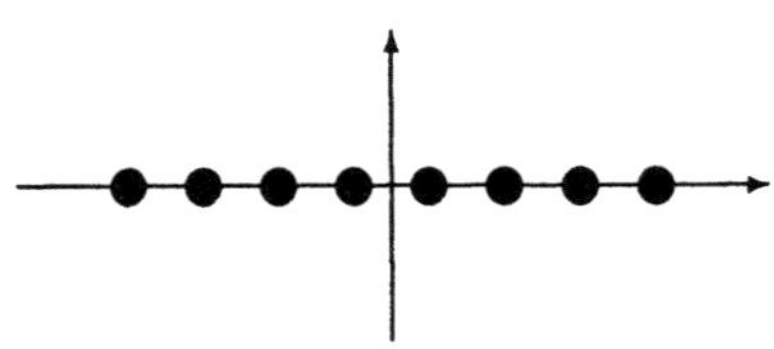

8-AM Constellation

The 8-PSK constellation can be viewed as the image of the ring $\mathbb{Z}_8$ of integers modulo 8 under the mapping

$$\varphi:\ \mathbb{Z}_8 \longrightarrow \mathbb{C}, \qquad t \longmapsto e^{\frac{2\pi i t}{8}}.$$

The 8-AM constellation can be viewed as the image under the mapping

$$\psi:\ \{0,1,\ldots,7\} \longrightarrow \mathbb{C}, \qquad t \longmapsto \frac{2t}{7} - 1.$$

In general a modulation $\varphi : R \longrightarrow \mathbb{C}$ induces an embedding of the code $\mathcal{C} \subset (R^n)^{\mathbb{Z}_+}$ into the space $(\mathbb{C}^n)^{\mathbb{Z}_+}$. If the transmission error has the statistics of additive white Gaussian noise (AWGN) as it is approximately the case in many communication systems then the decoding problem asks for the minimization of the error

$$\text{error} := \min_{v \in \mathcal{C}} \sum_{t \in \mathbb{Z}_+} \|v_t - \hat{v}_t\|^2. \tag{38.6}$$

As in Section 3 $\hat{v} = \{\hat{v}_t\}_{t \in \mathbb{Z}_+} \subset (\mathbb{C}^n)^{\mathbb{Z}_+}$ denotes the received message sequence. In contrast to (38.3) the Hamming distance is this time replaced with the Euclidean distance to the square. Decoding can always be achieved as long as the error is less than 1/2 of the value

$$d^2_{\min} := \min_{\substack{u,v \in \mathcal{C} \\ u \neq v}} \sum_{t \in \mathbb{Z}_+} \|u_t - v_t\|^2. \tag{38.7}$$

At first sight it seems that the problem of decoding on the Gaussian channel is covered by algorithms available in the systems literature such as e.g. [1, 7, 8]. Unfortunately there is a distinct difference: Although the code $\mathcal{C} \subset (R^n)^{\mathbb{Z}_+}$ is by definition R-linear it is obviously not true that the embedding into the sequence space $(\mathbb{C}^n)^{\mathbb{Z}_+}$ results into a $\mathbb{C}$-linear subspace.

In practice decoding is usually performed (as on the symmetric channel) using the Viterbi decoding algorithm. Once again this algorithm is limited to codes of fairly small McMillan degrees because of complexity considerations.

The problems presented in sections 3 and 4 seem to be hard in full generality. Indeed the problem contain the general problem of decoding linear block codes (transmitted over the symmetric or over the Gaussian channel) as a special instance. What seems to be feasible however is the construction of special classes of convolutional codes which come with efficient decoding algorithms. A step in this direction was done in [10]. It is the authors believe that progress towards the solution of above problem has significant implications for the way data is encoded and transmitted through various noisy communication channels. It would be a significant progress if any of the algorithms developed in the systems literature could be adapted to the problems presented in this paper.

Acknowledgments

The author would like to thank Roxana Smarandache and Oscar Takeshita for helpful discussions during the preparation of this paper.

5 References

[1] C. G. Goodwin and K. S. Sin. *Adaptive Filtering Prediction and Control.* Prentice Hall, 1984.

[2] S. Lin and D. Costello. *Error Control Coding: Fundamentals and Applications.* Prentice-Hall, Englewood Cliffs, NJ, 1983.

[3] H. A. Loeliger and T. Mittelholzer. Convolution codes over groups. *IEEE Trans. Inform. Theory*, 42(6):1660–1686, 1996.

[4] F. J. MacWilliams and N. J.A. Sloane. *The Theory of Error-Correcting Codes.* North Holland, Amsterdam, 1977.

[5] J. L. Massey and M. K. Sain. Codes, automata, and continuous systems: Explicit interconnections. *IEEE Trans. Automat. Contr.*, AC-12(6):644–650, 1967.

[6] Ph. Piret. *Convolutional Codes, an Algebraic Approach.* MIT Press, Cambridge, MA, 1988.

[7] B. Roorda. Algorithms for global total least squares modelling of finite multivariable time series. *Automatica J. IFAC*, 31(3):391–404, 1995.

[8] B. Roorda and C. Heij. Global total least squares modeling of multivariable time series. *IEEE Trans. Automat. Control*, 40(1):50–63, 1995.

[9] J. Rosenthal. Some interesting problems in systems theory which are of fundamental importance in coding theory. In *Proc. of the 36th IEEE Conference on Decision and Control*, pages 4574–4579, San Diego, California, 1997.

[10] J. Rosenthal. An algebraic decoding algorithm for convolutional codes. Preprint, January 1998.

[11] J. Rosenthal, J. M. Schumacher, and E.V. York. On behaviors and convolutional codes. *IEEE Trans. Inform. Theory*, 42(6):1881–1891, 1996.

[12] M.E. Valcher and E. Fornasini. On 2D finite support convolutional codes: an algebraic approach. *Multidim. Sys. and Sign. Proc.*, 5:231–243, 1994.

[13] S.B. Wicker. *Error Control Systems for Digital Communication and Storage.* Prentice Hall, New Jersey, 1995.

[14] J. C. Willems. Paradigms and puzzles in the theory of dynamical systems. *IEEE Trans. Automat. Control*, AC-36(3):259–294, 1991.

[15] S.G. Wilson. *Digital Modulation and Coding.* Prentice Hall, New Jersey, 1996.

[16] E.V. York. *Algebraic Description and Construction of Error Correcting Codes, a Systems Theory Point of View.* PhD thesis, University of Notre Dame, 1997.Available at
http://www.nd.edu/˜rosen/preprints.html.

39
Re-initialization in discontinuous systems

J.M. Schumacher

CWI
P. O. Box 94079
1090 GB Amsterdam
THE NETHERLANDS
Hans.Schumacher@cwi.nl

1 Description of the problem

Let an implicit system of differential equations be given in the form

$$f(x(t), \dot{x}(t)) = 0 \tag{39.1}$$

where x is an n-dimensional vector and f is a smooth mapping from $\mathbb{R}^{2n}$ to $\mathbb{R}^n$. More generally x may take values in a differentiable manifold and f may be defined on the tangent bundle. Let us first consider the case that f is linear, so that the equation (39.1) may be written as

$$E\dot{x}(t) = Fx(t) \tag{39.2}$$

where both E and F are matrices of size $n \times n$. To the pair of matrices (E, F) one may associate two subspaces of $\mathbb{R}^n$, in the following way. Denote by $V(E, F)$ the limit of the sequence defined by

$$V^0 = \mathbb{R}^n, \quad V^{j+1} = F^{-1}EV^j \quad (j = 0, 1, \ldots). \tag{39.3}$$

Alternatively, $V(E, F)$ may be defined as the largest element of the set of all subspaces V that satisfy $FV \subset EV$. Secondly, denote by $T(E, F)$ the limit of the sequence defined by

$$T^0 = \{0\}, \quad T^{j+1} = E^{-1}FT^j \quad (j = 0, 1, \ldots). \tag{39.4}$$

It is equivalent to say that $T(E, F)$ is the smallest element of the set of all subspaces T such that $T \supset E^{-1}FT$. We can now state the following well-known theorem.

Theorem 1 *For a pair of square real matrices (E, F), the following statements are equivalent:*

(i) *the polynomial* $\det(sE - F)$ *is nonzero;*

(ii) *the subspaces* $V(E,F)$ *and* $T(E,F)$ *form a direct sum decomposition of* $\mathbb{R}^n$*;*

(iii) *the differential equation (39.2) is well-posed in the sense that for each* $x_0 \in \mathbb{R}^n$ *there is at most one solution of (39.2) passing through* x_0.

Actually more precise statements could be made; in particular the "consistent subspace", that is, the space of all x_0 through which a smooth solution of (39.2) passes, is equal to $V(E,F)$, and the subspace $T(E,F)$ has an interpretation in terms of impulsive solutions to (39.2) [14, 11].
The theorem connects statements of three different types: an *algebraic* one, a *geometric* one, and a *dynamic* one (relating to differential equations). The question that we are interested in is now the following: *to what extent do there exist nonlinear generalizations of Thm. 1?* So, for a nonlinear implicit system of the form (39.1), we would like to know whether well-posedness can be connected to an algebraic criterion as in (i) above, and in particular (as motivated below) under what conditions there is an associated state space decomposition similar to (ii).
In the nonlinear case as given by (39.1) we can still define the *consistent manifold* as the set of all points x through which a smooth solution of (39.1) passes. Let us also define the *constraint manifold* as the set of all x for which there is a v such that $f(x,v) = 0$. Note that, in the linear case, the constraint manifold is exactly the subspace V^1 in the sequence defined by (39.3).

2 Motivation of the problem

Consider a linear system with scalar inputs and outputs

$$\dot{x}(t) = Ax(t) + Bu(t) \tag{39.5}$$

$$y(t) = Cx(t) \tag{39.6}$$

together with inequality constraints of the following form:

$$\text{for all } t\text{:}\quad y(t) \geq 0, \quad u(t) \geq 0, \quad \text{and} \quad y(t) = 0 \text{ or } u(t) = 0. \tag{39.7}$$

These could for instance be the equations of a linear electrical network with an ideal diode; also, equations of the same form arise from an application of Pontryagin's maximum principle to a linear-quadratic problem with a single linear inequality constraint. In the situation in which the constraint is active ($y = 0$), the system dynamics is given by

$$\frac{d}{dt}\begin{bmatrix} I & 0 \\ 0 & 0 \end{bmatrix}\begin{bmatrix} x \\ u \end{bmatrix}(t) = \begin{bmatrix} A & B \\ C & 0 \end{bmatrix}\begin{bmatrix} x \\ u \end{bmatrix}(t). \tag{39.8}$$

This is an equation of the form (39.2). If we assume that the system is well-posed, the conditions of the theorem mentioned above hold and so the consistent subspace is $V(E, F)$ with

$$E = \begin{bmatrix} I & 0 \\ 0 & 0 \end{bmatrix}, \quad F = \begin{bmatrix} A & B \\ C & 0 \end{bmatrix}. \tag{39.9}$$

Note that while it is natural to let a change from unconstrained to constrained mode take place only at times t at which both $y(t)$ and $u(t)$ vanish, the ensuing conditions $x \in \ker C$ and $u = 0$ may not be enough to guarantee that the vector $\left[\begin{smallmatrix} x \\ u \end{smallmatrix}\right]$ is in $V(E, F)$. However we are always able to project a given vector $\left[\begin{smallmatrix} x \\ u \end{smallmatrix}\right]$ onto $V(E, F)$ along the complementary subspace $T(E, F)$, and doing so is in fact supported by the interpretation of the latter space as the space of jump directions. The same recipe can be given in the case of systems multiple inputs and outputs connected by "complementarity conditions" of the form (39.7) (taken componentwise). Of course in the multivariable case there are more possible combinations of active and inactive constraints; a system with k constraints has 2^k modes. In this way it becomes possible to give an unequivocal specification of what is to be understood by a solution of a system of the form (39.5–39.7). The dynamical systems that are obtained in this way have been called *linear complementarity systems* [15]. Surely, to define the concept of a solution is one thing and to prove existence and uniqueness of solutions is another; sufficient conditions for well-posedness in terms of the classical linear input-output system (39.5) are given in [15].
Another example of a class of discontinuous dynamical systems for which a complete specification of the dynamics can be given is provided by mechanical systems subject to inequality constraints on the configuration variables; here the projection rule as defined by Moreau [25] can be used for re-initialization. In the class of general nonlinear systems, the re-initialization problem becomes vacuous if the consistent manifold coincides with the constraint manifold, so that the constraint only become active in situations in which the state is already in the consistent manifold corresponding to the relevant constraints, and state trajectories can be required to be continuous. Otherwise however there is a nontrivial re-initialization problem to solve, which will require a decomposition of the constraint manifold into the consistent manifold and an associated foliation — that is, a decomposition similar to the one in the theorem above.

3 History and related results

Theorem 1 above goes back 'essentially' to the 19th century, at least as far as the equivalence between parts (i) and (ii) is concerned, since the implication from (i) to (ii) can be read off from the Weierstrass canonical form for regular matrix pencils [32], and the implication from (ii) to (i) follows from the canonical form for singular matrix pencils given by Kronecker [20]. Weierstrass and Kronecker did not explicitly use the algorithms (39.3) and (39.4); these occur perhaps for the first time in a paper of 1946 by Dieudonné [6]. Gantmacher [9] presented a derivation of the Kronecker canonical form that is much simplified with respect to the original version, and also pointed out the implications of the canonical form for systems of implicit linear differential equations with constant coefficients. For systems in the special form (39.9) the algorithms (39.3) and (39.4) take on a special form as well, under which they have become well-known in control theory through the work of in particular Wonham [33], Morse [26], and Basile and Marro [1].

Implicit systems of nonlinear equations occur naturally in the modeling of physical phenomena, and in recent years there has been great interest in the development of numerical methods for solving such systems, see for instance [2, 12]. Implicit systems are categorized in classes of increasing difficulty of numerical solution by a quantity known as the "index". There are actually various definitions of this notion (cf. [5] for a discussion of several proposals), but roughly speaking the index is related to the number of "hidden equations" that are not explicitly formulated in the model equations but that follow from them by differentiation. The index can also be seen as a measure of the difference between the dimension of the consistent manifold and the dimension of the constraint manifold, so that problems that cause numerical difficulties are also problems that cause difficulties with re-initialization. One may surmise that this connection is not accidental; indeed the index reduction scheme proposed by Gear [10] defines a re-initialization mapping at least in a neighborhood of the consistent manifold, cf. the remark at the end of the cited paper.

It would appear therefore that the large literature on the index of implicit nonlinear systems will be of relevance to the re-initialization problem. Unfortunately this literature gives mixed clues; as already noted there are several definitions of the index which can give widely different results [5]. A case can be made for the claim that a study of the index should be based on methods from differential algebra. Steps in this direction have been taken by Fliess and co-authors [8, 7]. Prolongations (corresponding to 'repeated differentiation') have been used by Fliess *et al.* as well, and they have obtained a nonlinear version of the equivalence between (i) and (iii) in terms of differential functional independence [7]. See also the work by Le Vey [22] who develops connections to the formal theory of partial differential equations. The prolongation method is related to the use of 'dummy derivatives' in numerical treatments of implicit systems [24]. As an alternative to prolongation one may attempt to use more geometrically oriented methods as would be suggested by the linear version (39.4). A nonlinear variant of (39.4) has been proposed by Isidori *et al.* [18] although not at the level of generality of (39.1); moreover in some examples the algorithm of

[18] does not quite give the desired results, see [30, Remark 5.8]. Nonlinear versions of the algorithm (39.3) are well-known [17, 27, 28].
The linear case also suggests a connection to what is known in linear system theory as the "structure at infinity" (see for instance [21]). Indeed, the index of the linear system (39.2) can be defined as one plus the order of the pole at infinity of the rational matrix function $(sE-F)^{-1}$. Actually there are in general several poles at infinity, as indicated by the Smith-MacMillan form at infinity [13, 19], and this suggests that the index should be defined as a vector rather than as a single number. It has been shown by Bujakiewicz [4] that this extension is actually important for numerical purposes. Above it has been noted already that the subspace $T(E,F)$ has an interpretation in terms of impulsive solutions of the equation (39.2), which then needs to be interpreted in a generalized sense. This suggests that 'solving' (39.1) with an inconsistent initial condition (which is a possible interpretation of the re-initialization problem) will bring in the theory of impulsive solutions of nonlinear differential equations, which is a subject of its own. It may be expected that certain commutativity conditions on vector fields will be needed to ensure that solutions are well-defined; this is in line with the fact that involutivity conditions are often found to play an important role in applications of the nonlinear variant of (39.4) as given in [18], see for instance [31]. If such conditions are not satisfied it may happen that the re-initialization problem is not well-posed and one will either have to reformulate the model or accept a certain degree of indeterminism.

4 References

[1] G. Basile and G. Marro. *Controlled and Conditioned Invariants in Linear System Theory.* Prentice Hall, Englewood Cliffs, NJ, 1992.

[2] K. E. Brenan, S. L. Campbell, and L. R. Petzold. *Numerical Solution of Initial-Value Problems in Differential-Algebraic Equations.* North-Holland, Amsterdam, 1989.

[3] B. Brogliato. *Nonsmooth Impact Mechanics. Models, Dynamics and Control.* Lect. Notes Contr. Inform. Sci. 220. Springer, Berlin, 1996.

[4] P. Bujakiewicz. *Maximum weighted matching for high index differential algebraic equations.* PhD thesis, TU Delft, 1994.

[5] S.L. Campbell and C.W. Gear. The index of general nonlinear DAEs. *Numer. Math.*, 72:173–196, 1995.

[6] J. Dieudonné. Sur la réduction canonique des couples des matrices. *Bull. Soc. Math. France*, 74:130–146, 1946.

[7] M. Fliess, J. Lévine, Ph. Martin, and P. Rouchon. Index and decomposition of nonlinear implicit differential equations. In *Proc. IFAC*

Conference on System Structure and Control (Nantes, France, July 1995), pages 43–48, 1995.

[8] M. Fliess, J. Lévine, and P. Rouchon. Index of a general differential-algebraic implicit system. In H. Kimura and S. Kodama, editors, *Recent Advances in Mathematical Theory of Systems, Control, Networks and Signal Processing* (Proc. Int. Symp. MTNS-91, Kobe, Japan, June 1991), pages 289–294. Mita Press, 1992.

[9] F. R. Gantmacher. *The Theory of Matrices (Vol. II).* Chelsea, New York, 1959.

[10] C. W. Gear. Differential-algebraic equation index transformations. *SIAM J. Sci. Stat. Comput.*, 9:39–47, 1988.

[11] A. H. W. Geerts and J. M. Schumacher. Impulsive-smooth behavior in multimode systems. Part I: State-space and polynomial representations. *Automatica*, 32:747–758, 1996.

[12] E. Hairer and G. Wanner. *Solving Ordinary Differential Equations II: Stiff and Differential-Algebraic Problems.* Springer, Berlin, 1991.

[13] M. L. J. Hautus. The formal Laplace transform for smooth linear systems. In G. Marchesini and S.K. Mitter, editors, *Mathematical Systems Theory*, Lect. Notes Econ. Math. Syst. 131, pages 29–47. Springer, New York, 1976.

[14] M. L. J. Hautus and L. M. Silverman. System structure and singular control. *Lin. Alg. Appl.*, 50:369–402, 1983.

[15] W. P. M. H. Heemels, J. M. Schumacher, and S. Weiland. Linear complementarity systems. Internal Report 97 I/01, Measurement and Control Systems, Dept. of EE, Eindhoven Univ. of Technol., July 1997. `http://www.cwi.nl/~jms/lcs.ps.Z`.

[16] W. P. M. H. Heemels, J. M. Schumacher, and S. Weiland. The rational complementarity problem. Internal Report 98 I/02, Measurement and Control Systems, Dept. of EE, Eindhoven Univ. of Technol., May 1998.
`http://www.cwi.nl/~jms/PUB/rcp.ps.Z`.

[17] A. Isidori. *Nonlinear Control Systems: An Introduction.* Lect. Notes Contr. Inf. Sci. Springer, Berlin, 1985.

[18] A. Isidori, A.J. Krener, C. Gori-Giorgi, and S. Monaco. Nonlinear decoupling via feedback: a differential-geometric approach. *IEEE Trans. Automat. Contr.*, AC-26:331–345, 1981.

[19] T. Kailath. *Linear Systems.* Prentice-Hall, Englewood Cliffs, N.J., 1980.

[20] L. Kronecker. Algebraische Reduction der Schaaren bilinearer Formen. *S.-B. Akad. Berlin*, pages 763–776, 1890.

[21] M. Kuijper and J. M. Schumacher. Input-output structure of linear differential/algebraic systems. *IEEE Trans. Automat. Contr.*, AC-38:404–414, 1993.

[22] G. Le Vey. Differential algebraic equations, a new look at the index. Rapport de recherche 2239, INRIA, Rennes, 1994.

[23] O. Maler, editor. *Hybrid and Real-Time Systems.* (Proc. Intern. Workshop HART'97, Grenoble, France, March 1997.) Lect. Notes Comp. Sci. 1201, Berlin, 1997. Springer.

[24] S.E. Mattson and G. Söderlind. A new technique for solving high-index differential-algebraic equations using dummy derivatives. In *Proc. 1992 IEEE Symp. on Computer-Aided Control System Design (CACSD '92)*, Napa, Ca., March 1992, pages 218–224, 1992.

[25] J. J. Moreau. Liaisons unilatérales sans frottement et chocs inélastiques. *C. R. Acad. Sc. Paris*, 296:1473–1476, 1983.

[26] A. S. Morse. Structural invariants of linear multivariable systems. *SIAM J. Control*, 11:446–465, 1973.

[27] H. Nijmeijer and A. J. van der Schaft. *Nonlinear Dynamical Control Systems.* Springer-Verlag, Berlin, 1990.

[28] P.J. Rabier and W.C. Rheinboldt. A geometric treatment of implicit differential-algebraic equations. *J. Diff. Eq.*, 109:110–146, 1994.

[29] A. J. van der Schaft and J. M. Schumacher. Complementarity modeling of hybrid systems. *IEEE Trans. Automat. Contr.*, AC-43:483–490, 1998.

[30] A. J. van der Schaft and J. M. Schumacher. The complementary-slackness class of hybrid systems. *Math. Contr. Signals Syst.*, 9:266–301, 1996.

[31] A.J. van der Schaft. On realization of nonlinear systems described by higher-order differential equations. *Math. Syst. Th.*, 19:239–275, 1987. Correction: *Math. Syst. Th.*, 20:305–306, 1987.

[32] K. Weierstrass. Zur Theorie der bilinearen und quadratischen Formen. *Monatsh. Akad. Wiss.*, pages 310–338, 1867.

[33] W. M. Wonham. *Linear Multivariable Control: A Geometric Approach.* Springer Verlag, Heidelberg, 2nd edition, 1979.

40
Control-Lyapunov functions

Eduardo D. Sontag

Department of Mathematics
Rutgers University
New Brunswick, NJ 08903
USA
sontag@control.rutgers.edu

1 Motivation and history of the problems

The main objective of control is to modify the behavior of a dynamical system, typically with the purpose of regulating certain variables or of tracking desired signals. Often, either *stability* of the closed-loop system is an explicit requirement, or else the problem can be recast in a form that involves stabilization (e.g., of an error signal). For linear systems, the associated problems can now be treated fairly satisfactorily, but in the nonlinear case the area is still far from being settled. Both of the late 1980s reports [9] and [18], with dealt with challenges and future directions for research in control theory, identified the problem of stabilization of finite-dimensional deterministic systems as one of the most important open problems in nonlinear control. We discuss some questions in this area.
Specifically, this chapter deals with systems of the following general form:

$$\dot{x}(t) = f(x(t), u(t)). \tag{40.1}$$

The states $x(t)$ take values in Euclidean space $\mathbb{R}^n$ and the controls $u(t)$ take values in $\mathbb{R}^m$. The map $f(x,u)$ is continuous, and is locally Lipschitz in x, uniformly for u in compacts. In addition, $f(0,0) = 0$, that is to say, the zero state is an equilibrium when no inputs are applied. By a *control* we mean a measurable function $u : [0, +\infty) \to \mathbb{R}^m$ which is locally essentially bounded (meaning that, for each $T > 0$ there is some compact subset $K \subseteq \mathbb{R}^m$ so that $u(t) \in K$ for a.a. $t \in [0,T]$). In general, we use the notation $x(t; x_0, u)$ to denote the solution of (40.1) at time $t \geq 0$, with initial condition x_0 and control u. The expression $x(t; x_0, u)$ is defined on some maximal interval $[0, t_{\max}(x_0, u))$.
A common approach for the stabilization of (40.1) to $x = 0$ relies on the use of abstract "energy" or "cost" functions which can be made to in-

finitesimally decrease in directions corresponding to possible controls. Let us review some basic definitions. A function $V : \mathbb{R}^n \to \mathbb{R}_{\geq 0}$ is said to be *positive* (definite) if

$$V(x) > 0 \quad \forall\, x \neq 0,\; V(0) = 0, \tag{40.2}$$

and it is *proper* if the sublevel set $\{x | V(x) \leq a\}$ is compact, for each $a > 0$. The function V is said to be *infinitesimally decreasing* if there exists a continuous positive function $W : \mathbb{R}^n \to \mathbb{R}_{\geq 0}$ such that, for each compact set $K \subset \mathbb{R}^n$, there exists a compact set $U \subset \mathbb{R}^m$ so that

$$\min_{u \in U} \langle \nabla V(x), f(x,u) \rangle \;\leq\; -W(x) \quad \forall\, x \in K\,. \tag{40.3}$$

Note that this implies, in particular, the Hamilton-Jacobi inequality

$$\sup_{x \in \mathbb{R}^n} \;\inf_{u \in \mathbb{R}^m} \;\langle \nabla V(x), f(x,u) \rangle + W(x) \;\leq\; 0\,. \tag{40.4}$$

Definition. A continuously differentiable function $V : \mathbb{R}^n \to \mathbb{R}_{\geq 0}$ is called a *differentiable control Lyapunov function* (CLF) if it is positive, proper, and infinitesimally decreasing.

From a numerical point of view, given the objective of approaching the state $x = 0$, the use of control-Lyapunov functions reduces the search for stabilizing inputs to the iterative solution of a static nonlinear programming problem: when at state ξ, find u such that $\min_{u \in U} \langle \nabla V(x), f(x,u) \rangle \;\leq\; -W(x)$. This paradigm underlies the optimal control approach of Bellman, "artificial intelligence" techniques based on position evaluations in games and "critics" in learning programs, and several "neural-network" approaches to control.

Mathematically, the main implication of the existence of a CLF is null-asymptotic controllability. This means that for each initial state ξ there is some control function $u(\cdot)$ which steers the state ξ asymptotically to the origin, while not producing large excursions. More precisely:

Definition. The system (40.1) is *(globally) null-asymptotically controllable* if:

1. (**attractiveness**) for each $x_0 \in \mathbb{R}^n$ there exists some control u such that the trajectory $x(t) = x(t; x_0, u)$ is defined for all $t \geq 0$ and $x(t) \to 0$ as $t \to +\infty$;
2. (**Lyapunov stability**) for each $\varepsilon > 0$ there exists $\delta > 0$ such that for each $x_0 \in \mathbb{R}^n$ with $|x_0| < \delta$ there is a control u as in 1. such that $\|x(t)\| < \varepsilon$ for all $t \geq 0$;
3. (**bounded controls**) there are a neighborhood X of 0 in $\mathbb{R}^n$, and a compact subset U of $\mathbb{R}^m$ such that, if the initial state x_0 in 2. satisfies also $x_0 \in X$, then the control in 2. can be chosen with $u(t) \in U$ for almost all t.

This is a natural generalization to control systems of the concept of uniform asymptotic stability of solutions of differential equations.

2 Past work and new questions

A natural question is: can one use a known CLF in order to effectively design control laws that achieve stabilization? For systems affine in controls, Artstein's Theorem says that the existence of a differentiable CLF implies (and is also implied by) the existence of a continuous feedback stabilizer, cf. [1], meaning a $k : \mathbb{R}^n \to \mathbb{R}^m$ with the property that the closed-loop system $\dot{x} = f(x, k(x))$ has the origin $x = 0$ as a globally asymptotically stable equilibrium, and k is continuous away from the origin. CLF-based feedback designs provide one of the main current approaches to nonlinear control, and are discussed in detail in several recent textbooks, including [11, 10, 16, 20]. More precisely, one can find *universal formulas* for the stabilizer, which depend analytically on the directional derivatives of the CLF V. For example, for systems with one input ($m = 1$) $\dot{x} = f(x, u) = g_0(x) + ug_1(x)$, one has that

$$k(x) \; := \; - \frac{a(x) + \sqrt{a(x)^2 + b(x)^4}}{b(x)} \qquad (0 \text{ if } b = 0)$$

stabilizes the system, where we are denoting $a(x) := \nabla V(x)\, g_0(x)$ and $b(x) := \nabla V(x)\, g_1(x)$. (This expression is analytic on $a(x), b(x)$ when $x \neq 0$; cf. [23, 20].)

Often, controls are restricted to lie in constrained sets $\mathbb{U} \subseteq \mathbb{R}^m$, for instance due to actuator saturation effects. As Artstein's Theorem holds for arbitrary convex input-value sets $\mathbb{U}$, it would be desirable for design purposes to have universal formulas that lead to controls satisfying the same constraints, under the assumption that the CLF satisfies condition (40.3) with controls in $\mathbb{U}$. Examples of such formulas are given in ([14]), for open balls in $\mathbb{R}^n$.

Problem. *Find universal formulas for* CLF *stabilization, for general (convex) control-value sets* $\mathbb{U}$.

The above considerations motivate an obvious fundamental theoretical question: is the existence of a differentiable CLF *equivalent* to null-asymptotic controllability?

When there are no controls, the equivalence between asymptotic controllability and the existence of CLF's amounts to the equivalence between (global) asymptotic stability of an equilibrium and the existence of classical Lyapunov functions, and in that case the answer to the question is positive, and even an infinitely differentiable CLF always exists, as shown in the fundamental contributions of Massera and Kurzweil [15, 12]. When

there are controls, but the system is linear, again the answer is positive, and there is a quadratic CLF, as discussed in linear systems textbooks, e.g. [20]. In general, however, the answer to the question is negative. There do exist systems which are null-asymptotically controllable yet for which there is no possible differentiable CLF. There are various ways to prove this negative result. One is as a corollary of Artstein's Theorem, since in general continuous stabilizers are known not to exist for asymptotically controllable systems, cf. [17, 24, 21, 3, 2, 20]. See also [19, 7, 13] for many further recent results regarding the connection between continuous stabilizability and the existence of differentiable CLF's.
Faced with the negative answer, a very attractive relaxation of the question involves dropping the requirement that V be differentiable, and reinterpreting the differential inequality (40.3) in a weak sense. There are many candidates for this interpretation: viscosity solutions, generalized differentials, proximal subgradients, and others. These different interpretations, for the purpose of this note, are equivalent, as discussed in [5]. We pick one of them, proximal subgradients, for concreteness. Specifically, we say that a *continuous function* V is *infinitesimally decreasing* if the same definition as earlier holds, except that we substitute (40.3) by:

$$\min_{u\in U} \langle \zeta, f(x,u)\rangle \;\leq\; -W(x) \quad \forall\, x \in K \;\; \forall\, \zeta \in \partial_{\mathrm{P}} V(x)\,, \tag{40.5}$$

where $\partial_{\mathrm{P}} V(x)$ is the proximal subdifferential of the function V at the point x. We say that a continuous $V : \mathbb{R}^n \to \mathbb{R}_{\geq 0}$ is a (continuous) CLF if it is positive, proper, and infinitesimally decreasing in this sense.
Recall that a vector $\zeta \in \mathbb{R}^n$ is a proximal subgradient of V at x if there exists some $\sigma > 0$ such that, for all y in some neighborhood of x,

$$V(y) \;\geq\; V(x) + \langle \zeta, y - x\rangle - \sigma \left\|y - x\right\|^2 .$$

That is, ζ is the gradient of a supporting quadratic function at x to the graph of V. The set of proximal subgradients of V at x (which may be empty) is $\partial_{\mathrm{P}} V(x)$. The use of proximal subgradients as substitutes for the gradient for a nondifferentiable function was originally proposed in nonsmooth analysis for the study of optimization problems, see [4]. It is also possible to use continuous CLF's as a basis for feedback design; see [6]. The following positive result holds:

Theorem. A system is globally null-asymptotically controllable if and only if there exists a continuous CLF for it.

This was proved in [22]. More precisely, it was stated there in terms of Dini derivatives; the translation into the far more elegant and powerful language of proximal subgradients was later remarked in [6].
Thus, there is always a continuous V, but not always a differentiable V. This leads us to state:

Problem. *What is the best degree of regularity that can be assured for a* CLF, *for any globally null-asymptotically controllable system?*

Specifically, one may ask if one can always find a CLF that is piecewise differentiable on a locally finite (away from the origin) partition of $\mathbb{R}^n$, so that universal formulas for feedback stabilization can be employed locally.

3 References

[1] Artstein, Z., "Stabilization with relaxed controls," *Nonlinear Analysis, Theory, Methods & Applications* **7**(1983): 1163-1173.

[2] Bacciotti, A., *Local Stabilizability of Nonlinear Control Systems*, World Scientific, London, 1991.

[3] Brockett, R.W., "Asymptotic stability and feedback stabilization," in *Differential Geometric Control theory* (R.W. Brockett, R.S. Millman, and H.J. Sussmann, eds.), Birkhauser, Boston, 1983, pp. 181-191.

[4] Clarke, F.H., *Methods of Dynamic and Nonsmooth Optimization.* Volume 57 of *CBMS-NSF Regional Conference Series in Applied Mathematics*, S.I.A.M., Philadelphia, 1989.

[5] Clarke, F.H., Yu.S. Ledyaev, R.J. Stern, and P. Wolenski, *Nonsmooth Analysis and Control Theory*, Springer-Verlag, New York, 1998.

[6] Clarke, F.H., Yu.S. Ledyaev, E.D. Sontag, and A.I. Subbotin, "Asymptotic stabilization implies feedback stabilization," *IEEE Trans. Automat. Control* **42**(1997): 1394-1407

[7] Coron, J.-M., and L. Rosier, "A relation between continuous time-varying and discontinuous feedback stabilization," *J. Math. Systems Estim. Control* **4** (1994), 67-84.

[8] Coron, J.M., L. Praly, and A. Teel, "Feedback stabilization of nonlinear systems: sufficient conditions and Lyapunov and input-output techniques," in *Trends in Control: A European Perspective* (A. Isidori, Ed.), Springer, London, 1995 (pp. 293-348).

[9] Fleming, W.H. et. al., *Future Directions in Control Theory: A Mathematical Perspective*, SIAM Publications, Philadelphia, 1988.

[10] Krstic, M. and H. Deng, *Stabilization of Uncertain Nonlinear Systems*, Springer-Verlag, London, 1998.

[11] Krstic, M., I. Kanellakopoulos, and P. Kokotovic, *Nonlinear and adaptive control design*, John Wiley & Sons, New York, 1995.

[12] Kurzweil, J., "On the reversibility of the first theorem of Lyapunov concerning the stability of motion," *Czechoslovak Math. J.* **5(80)** (1955), 382–398 (in Russian). English translation in: *Amer. Math. Soc. Translations, Series 2*, **24** (1956), 19-77.

[13] Ledyaev, Yu.S., and E.D. Sontag, "A Lyapunov characterization of robust stabilization," Proc. Automatic Control Conference, Philadelphia, June 1998.

[14] Lin, Y., and E.D. Sontag, "Control-Lyapunov universal formulae for restricted inputs," *Control: Theory and Advanced Technology* **10**(1995): 1981-2004.

[15] Massera, J.L., "Contributions to stability theory," *Annals of Math* **64** (1956): 182-206. Erratum in *Annals of Math* **68**(1958): 202.

[16] Isidori, A., *Nonlinear Control Systems: An Introduction*, Springer-Verlag, Berlin, third ed., 1995.

[17] Jurdjevic, V. and J.P. Quinn, "Controllability and stability," *J.of Diff.Eqs.* **28** (1978): 381-389.

[18] Levis, A.H., et. al., "Challenges to control: A collective view. Report from the workshop held at the University of Santa Clara on September 18-19, 1986," *IEEE Trans. Autom. Control* **32**(1987): 275-285.

[19] Ryan, E.P., "On Brockett's condition for smooth stabilizability and its necessity in a context of nonsmooth feedback," *SIAM J. Control Optim.* **32** (1994) 1597–1604.

[20] Sontag, E.D., *Mathematical Control Theory: Deterministic Finite Dimensional Systems. Second Edition*, Springer, New York, 1998.

[21] Sontag, E.D., and H.J. Sussmann, "Remarks on continuous feedback," *Proc. IEEE Conf. Decision and Control, Albuquerque, Dec.1980*, pp. 916-921.

[22] Sontag, E.D., "A Lyapunov-like characterization of asymptotic controllability," *SIAM J. Control & Opt.* **21**(1983): 462-471.

[23] Sontag, E.D., "A 'universal' construction of Artstein's theorem on nonlinear stabilization," *Systems and Control Letters*, **13**(1989): 117-123.

[24] Sussmann, H.J., "Subanalytic sets and feedback control," *J. Diff. Eqs.* **31**(1979): 31-52.

41
Spectral Nevanlinna-Pick Interpolation

Allen R. Tannenbaum
Department of Electrical and Computer Engineering
University of Minnesota
Minneapolis, MN 55455
USA
tannenba@ece.umn.edu

1 Description of the problem

Let $\mathcal{E}$ be a finite dimensional vector space, let D denote the unit disc, let $z_1, \ldots, z_n \in D$ be mutually distinct, and let $F_1, \ldots, F_n \in B(\mathcal{E})$, where $B(\mathcal{E})$, denotes the space of bounded linear operators on $\mathcal{E}$. For a bounded analytic function $F : D \to B(\mathcal{E})$, let $\|F\|_{\text{sp}}$ be the *spectral radius* defined by

$$\|F\|_{\text{sp}} := \sup\{\|F(z)\|_{\text{sp}} : \quad z \in D\},$$

where $\|F(z)\|_{\text{sp}}$ is the spectral radius of the operator $F(z)$ (= the absolute value of the eigenvalue of largest magnitude of $F(z)$).
We are interested in problems of the following type:

1. Find necessary and sufficient conditions for the existence of an analytic function $F : D \to B(\mathcal{E})$ with $\|F\|_{\text{sp}} \leq 1$ such that

$$F(z_j) = F_j$$

   for $j = 1, \ldots, n$.

2. If a solution exists, when does there exist an optimal solution?

3. Characterize the set of optimal solutions.

In contrast, to classical norm-bounded interpolation, it is known that there may not be an optimal solution for the above spectral interpolation problem [1, 2].

2 Motivation and history of the problem

Over the past 15 years, there have appeared a large number of papers concerned with the application of Nevanlinna-Pick interpolation theory to various problems in control, most notably the area of H^∞-optimization theory pioneered by George Zames in his seminal work [13]. Interpolation by bounded analytic functions in the disc is a topic with a rich mathematical history and wide-ranging applications [7]. Indeed, here the areas of complex analysis, operator theory, and hyperbolic geometry intersect creating a rich synthesis and a research subject with its own flavor and techniques.

Now the connection of interpolation theory to problems involving LTI, finite dimensional plants is very simple. Namely one may easily show that the question of internal stabilization amounts to a Lagrange-type interpolation problem for such plants [4, 10]. The way then that H^∞ optimal sensitivity reduces to a Nevanlinna-Pick interpolation problem is that one imposes a bound on the norm of the corresponding interpolating functions [4, 5, 13]. For SISO systems gain margin optimization works in the same way in that via a conformal equivalence, we can transform the question of finding an interpolating function whose range is a simply-connected subset of the complex plane to one whose range is the unit disc and hence once more derive an interpolation problem of the Nevanlinna-Pick kind [10]. For MIMO systems, one can show that if one wants to play the same game with multivariate generalizations of gain margin, one derives interpolation not with a norm constraint but with a spectral radius constraint of the type described above [11, 12].

3 Available results and Extensions

The main available results are based on spectral generalizations of the classical commutant lifting theorem [9, 8]. This theorem relates the optimal solution of a Nevanlinna-Pick interpolation problem to the norm of an associated Hankel operator.

In a similar manner, the optimal solution of the spectral problem (when it exists) can be related to a generalized spectral radius defined in the following manner: Let $\mathcal{H}$ be a complex Hilbert space. Given $A, T \in B(\mathcal{H})$, we let

$$\{T\}' := \{A \in B(\mathcal{H}) : AT = TA\}.$$

Then we set

$$\rho_T(A) := \inf\{\|MAM^{-1}\| : M \text{ is invertible and } M \in \{T\}'\}.$$

We call $\rho_T(A)$ the *T-spectral radius of A*. The optimal spectral interpolant, then can be formulated in terms of $\rho_T(A)$. This however, does not give the conditions when in fact such an optimal interpolant exists.

There are moreover structured extensions of these results relating to the structured singular vlaue and μ–synthesis. See [3].

4 Partial solution

There are some partial solutions characterizing the optimal solutions. For example, we have the following necessary condition from [2]:

Let $z_1, z_2, \ldots, z_n \in D$ be distinct points, and $F_1, F_2, \ldots, F_n$ be given $N \times N$ complex matrices. Assume that the spectral Nevanlinna-Pick problem defined by this data has an optimal solution F such that $\|F\|_{\mathrm{sp}} = \|F\|_\infty = 1$. Then $\|F(\zeta)\|_{\mathrm{sp}} = 1$ for almost every $\zeta \in \partial D$.

Precise conditions guaranteeing the existence of an optimal solution are still lacking however.

5 References

[1] H. Bercovici, C. Foias, and A. Tannenbaum, "On the optimal solutions in spectral commutant lifting theory," *Journal of Functional Analysis* **101** (1991), pp. 38–49.

[2] H. Bercovici, C. Foias, and A. Tannenbaum, "A spectral commutant lifting theorem," *Trans. AMS* **325** (1991), pp. 741–763

[3] H. Bercovici, C. Foias, and A. Tannenbaum, "The structured singular value for linear input/output operators," *SIAM J. Control and Optimization* **34** (1996), pp. 1392–1404.

[4] J. C. Doyle, B. Francis, and A. Tannenbaum, *Feedback Control Theory*, McMillan, New York, 1991.

[5] B. Francis, *A Course in H^∞ Control Theory*, Lecture Notes in Control and Information Sciences **88**, Springer Verlag, 1987.

[6] B. Francis and A. Tannenbaum, "Generalized interpolation theory in control," *Mathematical Intelligencer* **10** (1988), pp. 48–53.

[7] J. Garnett, *Bounded Analytic Functions*, Academic Press, New York, 1981.

[8] D. Sarason, "Generalized interpolation in H^∞," *Transactions of the AMS* **127** (1967), pp. 179–203.

[9] B. Sz.-Nagy and C. Foias, *Harmonic Analysis of Operators on Hilbert Space*, North Holland, Amsterdam, 1970.

[10] A. Tannenbaum, *Invariance and System Theory: Algebraic and Geometric Aspects*, Lecture Notes in Mathematics **845**, Springer-Verlag, 1981.

[11] A. Tannenbaum, "Spectral Nevanlinna-Pick interpolation theory," *Proc. of 26th IEEE Conference on Decision and Control*, Los Angeles, California, December 1987, pp. 1635–1638.

[12] A. Tannenbaum, "On the multivariable gain margin problem," *Automatica* **22** (1986), pp. 381–384.

[13] G. Zames, "Feedback and optimal sensitivity: model reference transformations, multiplicative seminorms, and approximate inverses," *IEEE Trans. Auto. Control* **AC-26** (1981), pp. 301–320.

42

Phase-sensitive structured singular value

André L. Tits* and V. Balakrishnan**

*Department of Electrical Engineering and Institute for Systems Research
University of Maryland
College Park, MD 20742
USA
andre@isr.umd.edu

**School of Electrical and Computer Engineering
Purdue University
West Lafayette, IN 47907-1285
USA
ragu@ecn.purdue.edu

1 Description of the problem in the simplest case

Given $\Gamma \in \mathbf{C}^{n\times n}$ with $\Gamma + \Gamma^* \geq 0$, define the phase $\Phi(\Gamma)$ of Γ by

$$\Phi(\Gamma) = \cot^{-1}\left(\sup\left\{b:\ \Gamma + \Gamma^* - \frac{\beta}{\mathrm{j}}(\Gamma - \Gamma^*) \geq 0\ \forall\beta \in \{-b, b\}\right\}\right).$$

For $\theta \in [0, \pi/2]$, let

$$\mathbf{\Gamma}_\theta = \{\Gamma \in \mathbf{C}^{n\times n} : \Gamma + \Gamma^* \geq 0,\ \Phi(\Gamma) \leq \theta\},$$

and

$$\mathbf{BR\Delta}_\theta = \{\Delta \in \mathbf{RH}_\infty^{n\times n} :\ \|\Delta\|_\infty \leq 1,\ \Delta(\mathrm{j}\omega) \in \mathbf{\Gamma}_\theta \quad \forall\omega \in \mathbf{R}\},$$

where $\mathbf{RH}_\infty$ denotes the set of real-rational functions in $\mathbf{H}_\infty$. Following [1, 2], define the *phase-sensitive structured singular value* of $M \in \mathbf{C}^{n\times n}$ for phase θ by

$$\mu_\theta(M) = \left(\inf\{\overline{\sigma}(\Gamma) : \Gamma \in \mathbf{\Gamma}_\theta, \det(I + \Gamma M) = 0\}\right)^{-1}$$

if $\det(I + \Gamma M) = 0$ for some $\Gamma \in \mathbf{\Gamma}_\theta$, and $\mu_\theta(M) = 0$ otherwise. ($\overline{\sigma}(\Gamma)$ stands for the maximum singular value of Γ.) Also define

$$\hat{\mu}_\theta(M)$$

$$= \inf \left\{ \gamma : \begin{array}{l} r(M^*M - \gamma^2 I) - ((1+\mathrm{j}\beta)M + ((1+\mathrm{j}\beta)M)^*) < 0, \\ \gamma > 0,\ r > 0,\ \beta \in [-\cot\theta, \cot\theta] \text{ when } \theta > 0 \end{array} \right\}. \tag{42.1}$$

For each of the following two statements, we are interested in determining conditions under which the statement holds.

1. For $P \in \mathbf{RH}_\infty^{n\times n}$, $\theta \in [0, \pi/2]$:

$$\sup_{\omega \in \mathbf{R}\cup\infty} \mu_\theta(P(\mathrm{j}\omega)) < 1 \tag{42.2}$$

if, and only if, $(I + \Delta P)^{-1} \in \mathbf{H}_\infty$ for all $\Delta \in \mathbf{BR}\boldsymbol{\Delta}_\theta$ and

$$\sup_{\Delta \in \mathbf{BR}\boldsymbol{\Delta}_\theta} \|(I + \Delta P)^{-1}\|_\infty < \infty.$$

2. For $M \in \mathbf{C}^{n\times n}$, $\theta \in [0, \pi/2]$:

$$\mu_\theta(M) = \hat{\mu}_\theta(M).$$

2 Motivations

The phase-sensitive structured singular value was introduced in [1, 2] as a tool for the analysis of robust stability when, in addition to a magnitude bound, certain phase information is available concerning the uncertainty; see also [3]. Indeed, it can be checked that a matrix Γ belongs to $\boldsymbol{\Gamma}_\theta$ if, and only if, its numerical range (field of values) is contained in a sector of aperture 2θ about the positive real axis. For example, $\theta = \pi/2$ corresponds to the case when the uncertainty is known to be passive. For given M and θ, $\hat{\mu}_\theta(M)$ can be evaluated by solving a linear matrix inequality GEVP (generalized eigenvalue minimization problem, [2, 4]). It is easily shown that $\hat{\mu}_\theta(M)$ is always an upper bound to $\mu_\theta(M)$. The first of the statements above, when it holds, is a "small-μ theorem." The second statement, when it holds, implies that exact computation of $\mu_\theta(M)$ is tractable.

3 More general formulation

The case of block-diagonal structures, with different (possibly frequency-dependent) phase bounds on each block, is of interest as well. Assume ℓ blocks of size $k_1, \ldots, k_\ell$, let $\Theta = (\theta_1, \ldots, \theta_\ell)$ with $\theta_i \in [0, \pi/2]$ and define $\boldsymbol{\Gamma}_\Theta$ and $\mu_\Theta(M)$ in the obvious way. Then

$$\hat{\mu}_\Theta(M) = \inf \left\{ \gamma : \begin{array}{l} M^*RM - \gamma^2 R - S(I + \mathrm{j}B)M - M^*(I - \mathrm{j}B)S < 0, \\ \gamma > 0,\ R,\ S \in \mathcal{S},\ B \in \mathcal{B} \end{array} \right\},$$

where

$$\mathcal{S} = \{\mathrm{diag}(s_1 I_{k_1}, \ldots, s_\ell I_{k_\ell}) \ : \ s_i > 0\},$$

and

$$\mathcal{B} = \{\mathrm{diag}(\beta_1 I_{k_1}, \ldots, \beta_\ell I_{k_\ell}) : \beta_i \in [-\cot\theta_i, \cot\theta_i] \text{ when } \theta_i > 0\},$$

and I_k is a $k \times k$ identity matrix. (Note that the unstructured case (1) corresponds to $\ell = 1$.)

4 Available results

Concerning Statement 1, the following is shown in [2]:

- Sufficiency of (42.2) always holds. Furthermore, the following weaker "small-μ theorem" always holds: (42.2) holds if, and only if, $(I + \Delta P)^{-1} \in \mathbf{H}_\infty$ for all $\Delta \in \mathbf{B\Delta}_\theta$ and

$$\sup_{\Delta \in \mathbf{B\Delta}_\theta} \|(I + \Delta P)^{-1}\|_\infty < \infty,$$

  where

$$\mathbf{B\Delta}_\theta = \{\Delta \in \mathbf{H}_\infty^{n\times n} : \ \|\Delta\|_\infty \leq 1, \ \Delta(\mathrm{j}\omega) \in \mathbf{\Gamma}_\theta \quad \forall\omega \in \mathbf{R}\}.$$

  Note that $\mathbf{B\Delta}_\theta$ is much larger than $\mathbf{BR\Delta}_\theta$, as it includes, among other transfer functions, all complex matrices in $\mathbf{\Gamma}_\theta$.

- In the scalar case (as well as in the diagonal uncertainty case), necessity also holds. The key is the following lemma, proved in [2]: Let $\theta \in (0, \pi/2]$, let $\hat{\omega} \in \mathbf{R} \setminus \{0\}$, and let $\gamma \in \mathbf{C}$ be such that $|\gamma| < 1$ and $|\phi(\gamma)| < \theta$. There exists $\delta \in \mathbf{RH}_\infty$ such that $\delta(\mathrm{j}\hat{\omega}) = \gamma$ and such that $\sup_{\omega\in\mathbf{R}} |\delta(\mathrm{j}\omega)| < 1$ and $\sup_{\omega\in\mathbf{R}} |\phi(\delta(\mathrm{j}\omega))| < \theta$.

It is easily verified that Statement 2 holds in the scalar case. Specifically, for $m \in \mathbf{C}$ and any $\theta \in [0, \pi/2]$, $\mu_\theta(m)$ and $\hat{\mu}_\theta(m)$ are both equal to $|m|$ when $|\phi(m)| \geq \pi - \theta$, and 0 otherwise, where $\phi(\cdot)$ is the phase taken in $(-\pi, \pi]$ and $\phi(0) = 0$.

5 References

[1] L. Lee, "Robustness Study of Systems with Phase-Informed Uncertainty," Ph.D. Dissertation, Department of Electrical Engineering, University of Maryland, College Park, MD 20742, 1992.

[2] A. L. Tits, V. Balakrishnan, and L. Lee, "Robustness Under Bounded Uncertainty with Phase Information," to appear in *IEEE Trans. Automatic Control*, November 1998.

[3] E. G. Eszter and C. V. Hollot, "An IQC for Uncertainty Satisfying Both Norm-bounded and Passivity Constraints," *Automatica*, 33(8), pp. 1545–1548 (1997).

[4] S. Boyd, L. El Ghaoui, E. Feron, and V. Balakrishnan. *Linear Matrix Inequalities in System and Control Theory*, volume 15 of *Studies in Applied Mathematics*. SIAM, Philadelphia, PA, June 1994.

43

Conservatism of the standard upper bound test: Is $\sup(\overline{\mu}/\mu)$ finite? Is it bounded by 2?

Onur Toker* and Bram de Jager**

*College of Computer Science and Engineering
King Fahd University of Petroleum and Minerals
P.O. Box 14
Dhahran 31261
SAUDIA ARABIA
onur@ccse.kfupm.edu.sa

**Faculty of Mechanical Engineering
Eindhoven University of Technology
P.O. Box 513
5600 MB Eindhoven
THE NETHERLANDS
jag@wfw.wtb.tue.nl

1 Introduction

This short note is about the conservatism of the standard upper bound, $\overline{\mu}$, relative to the complex structured singular value, μ. This problem is first formulated by John Doyle. To follow the present note, only mathematical definitions of μ and $\overline{\mu}$ are necessary. For a tutorial introduction about complex μ and its importance in system analysis and design, see [10]. If $M \in \mathbb{C}^{n\times n}$, then

$$\mu(M) := \sup\{\rho(M\Delta) \ : \ \Delta = \mathrm{diag}(\delta_1, \ldots, \delta_n), \ \delta_1, \ldots, \delta_n \in \mathbf{D}\},$$

$$\overline{\mu}(M) := \inf\{\overline{\sigma}(D^{-1}MD) \ : \ D = \mathrm{diag}(d_1, \ldots, d_n), \ d_1, \ldots, d_n > 0\}.$$

It is well-known that

$$\mu(M) \leq \overline{\mu}(M),$$

equality holds if $\dim(M) \leq 3$, and there exist matrices of dimension 4 for which the inequality is strict [10]. But, it is not known whether there exists a finite number $g \geq 1$ such that

$$\overline{\mu}(M) \leq g\, \mu(M).$$

All of the known examples show that for $g = 2$ the inequality is always satisfied, but there is no known formal proof. Establishing such an inequality will be of great value, because μ is $\mathcal{NP}$-hard to compute whereas $\overline{\mu}$ is efficiently computable, and inequalities like

$$\frac{1}{2}\overline{\mu}(M) \leq \mu(M) \leq \overline{\mu}(M),$$

would imply that $\overline{\mu}$ is a computationally attractive alternative for μ. This will provide a theoretical justification for the use of $\overline{\mu}$ instead of μ.
Shamma [13], Megretski [7, 8], and Poolla and Tikku [11] proved that $\overline{\mu}$ is to be used in "if and only if" tests for robustness against time varying perturbations and hence provided alternative theoretical justifications for its use instead of μ.
Numerical results show that $\overline{\mu}$ is a "reliable" estimate of μ, and we are interested in different ways of proving (or disproving) this result. Computation of complex μ is known to be $\mathcal{NP}$-hard [14, 16] (See [2, 9, 12] for related complexity results in the context of real and mixed μ), however computation of $\overline{\mu}$ reduces to a simple LMI problem [10] and hence can be solved quite easily [1]. More precisely, there exists an algorithm (i.e., the interior point algorithm) with computation time bounded by a polynomial function of the problem size [5, 1]. What is not known is the worst case $\overline{\mu}/\mu$ ratio. Megretski gave the following geometric characterization of this problem in [6] (See [3, 15, 4] for related results in the context of real and mixed μ). Let g_n be the minimum $g \in \mathbb{R}^+$ such that for all unit vectors $a_1, \ldots, a_n, b_1, \ldots, b_n \in \mathbb{C}^n$, there exists a unit vector $x \in \mathbb{C}^n$ with

$$g|x^* a_k| \geq |x^* b_k|, \quad k = 1, \ldots, n.$$

Then

$$g_n = \sup\left\{ \frac{\overline{\mu}(M)}{\mu(M)} \; : \; \mu(M) \neq 0, \quad \dim(M) = n \right\},$$

2 Mathematical statement of Problem 1:

Is there a $g \in \mathbb{R}^+$ such that, for all $n \in \mathbb{Z}^+$, for all unit vectors

$$a_1, \ldots, a_n, b_1, \ldots, b_n \in \mathbb{C}^n,$$

there exists a unit vector $x \in \mathbb{C}^n$ such that

$$g|x^* a_k| \geq |x^* b_k|, \quad k = 1, \ldots, n.$$

If such a finite g exists, can it be chosen less than or equal to 2?

3 Mathematical statement of Problem 2 (Main Problem):

Is the following statement true or false: For all $n \in \mathbb{Z}^+$, for all unit vectors $a_1, \ldots, a_n, b_1, \ldots, b_n \in \mathbb{C}^n$, there exists a $x \in \mathbb{C}^n$ such that

$$2|x^* a_k| \geq |x^* b_k|, \quad k = 1, \ldots, n.$$

4 References

[1] Boyd, S., L. El Ghaoui, E. Feron, V. Balakrishnan, *Linear Matrix Inequalities in System and Control Theory.* Philadelphia: Society for Industrial and Applied Mathematics, 1994.

[2] Braatz, R. D., P. M. Young, J. C. Doyle, and M. Morari, "Computational Complexity of μ Calculation," *IEEE Transactions on Automatic Control*, **39** (1994) pp. 1000-1002.

[3] Coxson, G. E., and C. L. DeMarco, "The computational complexity of approximating the minimal perturbation scaling to achieve instability in an interval matrix," *Math. of Control, Signals, and Systems*, **7** (1994) pp. 279–292.

[4] Fu, M., "The real structured singular value is hard to approximate," to appear in *IEEE Transactions on Automatic Control.*

[5] Garey, M., and D. Johnson, *Computers and Intractability: A Guide to the Theory of $\mathcal{NP}$-completeness*, San Francisco: W. H. Freeman, 1979.

[6] Megretski, A., "On the gap between structured singular values and their upper bounds," *Proc. 32nd Conference on Decision and Control*, San Antonio, TX, Dec. 1993, pp. 3461–3462.

[7] Megretski, A., "Necessary and sufficient conditions of stability: a multiloop generalization of the circle criterion," *IEEE Transactions on Automatic Control*, **38** (1993) pp. 753-756.

[8] Megretski, A., and S. Treil, "Power distribution inequalities in optimization and robustness of uncertain systems," *Journal of Math. Systems, Estimation, Control*, **3** (1993) pp. 301–319.

[9] Nemirovskii, A., "Several $\mathcal{NP}$-hard problems arising in robust stability analysis," *Math. of Control, Signals, and Systems,* **6** (1993) pp. 99–105.

[10] Packard, A., and J. C. Doyle, "The Complex Structured Singular Value," *Automatica,* **29** (1993) pp. 71–109.

[11] Poolla, K., and A. Tikku, "Robust Performance Against Time-Varying Structured Perturbations," *IEEE Transactions on Automatic Control*, **40** (1995) pp. 1589–1602.

[12] Poljak, S., and J. Rohn, "Checking robust nonsingularity is $\mathcal{NP}$-hard," *Math. of Control, Signals, and Systems*, **6** (1993) pp. 1–9.

[13] Shamma, J., "Robust Stability with Time Varying Structured Uncertainity," *IEEE Transactions on Automatic Control*, **39** (1994) pp. 714–724.

[14] Toker, O., and H. Ozbay, "On the Complexity of Purely Complex μ Computation and Related Problems in Multidimensional Systems," *IEEE Transactions on Automatic Control*, **43** (1998) pp. 409–414.

[15] Toker, O., "On the conservatism of the upper bound tests for structured singular value analysis," *Proc. 35th Conference on Decision and Control*, Kobe, Japan, Dec. 1996, pp. 1295–1300.

[16] Toker, O., and H. Ozbay, "Complexity issues in robust stability of linear delay-differential systems," *Math. of Control, Signals, and Systems*, **9** (1996) pp. 386–400.

44

When does the algebraic Riccati equation have a negative semi-definite solution?

Harry L. Trentelman

Research Institute for Mathematics and Computing Science
P.O. Box 800
9700 AV Groningen
THE NETHERLANDS
H.L.Trentelman@math.rug.nl

1 Introduction

In this contribution we want to draw the readers's attention to an open problem that concerns the existence of certain solutions to the algebraic Riccati equation. Since its introduction in control theory by Kalman in the beginning of the sixties, the algebraic Riccati equation has known an impressive range of applications, such as linear quadratic optimal control, stability theory, stochastic filtering and stochastic control, stochastic realization theory, the synthesis of linear passive networks, differential games, and H_∞ optimal control and robust stabilization. For an overview of the existing literature on the algebraic Riccati equation, we refer to [3].
In this note, we deal with the *existence* of real symmetric solutions to the algebraic Riccati equation, in particular with the existence of *negative semi-definite* solutions. It is well-known (see [9]) that the existence of a real symmetric solution to the algebraic Riccati equation is equivalent to a given frequency domain inequality along the imaginary axis. In [9], it was also stated that the existence of a *negative semi-definite solution* is equivalent to this frequency domain inequality holding for all complex numbers in the closed right half of the complex plane. Soon after the appearance of [9], a correction [10] appeared in which it was outlined that this statement is not correct, and that the frequency domain inequality in the closed right half plane is a necessary, but not sufficient condition for the existence of a negative semi-definite solution.
Since then, several attempts have been made to obtain a convenient necessary and sufficient frequency domain condition for the existence of a nega-

tive semi-definite solution. In this note we discuss some of these conditions. We also explain that we consider these conditions not to be satisfactory yet, and therefore we claim the problem of formulating a sensible frequency domain condition to be still an open problem.

2 The algebraic Riccati equation

Let $A \in \mathbf{R}^{n\times n}$ and $B \in \mathbf{R}^{n\times m}$ be such that (A, B) is a controllable pair. Also, let $Q \in \mathbf{R}^{n\times n}$ be such that $Q = Q^T$, let $S \in \mathbf{R}^{m\times n}$, and let $R \in \mathbf{R}^{m\times m}$ be such that $R > 0$. We deal with the following algebraic Riccati equation:

$$A^T K + KA + Q - (KB + S^T)R^{-1}(B^T K + S) = 0 \qquad (44.1)$$

In order to give a frequency domain condition for the existence of a real symmetric solution K, we define the matrix of two-variable rational functions W in the indeterminates ζ and η by

$$\begin{aligned} W(\zeta,\eta) := R + B^T(\zeta I - A^T)^{-1}S^T + S(\eta I - A)^{-1}B \\ +B^T(\zeta I - A^T)^{-1}Q(\eta I - A)^{-1}B \end{aligned} \qquad (44.2)$$

which is often called the *Popov function* associated with (44.1). (see [7]). The coefficients of this matrix are quotients of real polynomials in the indeterminates ζ and η. With this two-variable rational matrix W we associate a (one-variable) rational matrix ∂W by defining $\partial W(\xi) := W(-\xi, \xi)$, i.e., obtained by taking $\zeta = -\xi$ and $\eta = \xi$. It was shown in [9] that the algebraic Riccati equation (44.1) has a real symmetric solution if and only if

$$\partial W(i\omega) \geq 0 \text{ for all } \omega \in \mathbf{R},\ i\omega \notin \sigma(A) \qquad (44.3)$$

In [9] it was claimed that the two-variable rational matrix W also provides the clue to the existence of a negative semi-definite solution. Indeed, [9], Theorem 4 states that (44.1) has a real symmetric solution $K \leq 0$ if and only if

$$W(\bar{\lambda}, \lambda) \geq 0 \text{ for } \Re e(\lambda) \geq 0,\ \lambda \notin \sigma(A) \qquad (44.4)$$

Unfortunately, as was noted in [10], this statement is not correct: in general the condition (44.4) is only a necessary condititition, but not a sufficient one. It is certainly interesting to note that for certain important special cases (44.4) does provide a necessary *and* sufficient condition. For one important special case this result is known as *the bounded real lemma.* This special case in concerned with the situation that $Q = -C^T C$, $S = 0$, and $R = I$. In that case, after a change of variable from K to $-K$, the Riccati equation becomes

$$A^T K + KA + C^T C + KBB^T K = 0 \qquad (44.5)$$

while we have $W(\zeta,\eta) = I - G^T(\zeta)G(\eta)$. The condition (44.4) then becomes

$$G^T(\bar{\lambda})G(\lambda) \leq I \text{ for } \Re e(\lambda) \geq 0,\ \lambda \notin \sigma(A)$$

which indeed is well-known to be equivalent to the existence of a solution $K \geq 0$ to the algebraic Riccati equation (44.5) (see, for example [2]). In [5], a number of additional special cases in which the frequency domain inequality (44.4) is a necessary and sufficient condition were established. We will not go into these special cases here. Instead, we want to discuss an alternative condition on the two-variable rational matrix W that was proven in [4] to be equivalent to the existence of a negative semi-definite solution to (44.1). In order to rederive this condition here, we make use of recent results in [12] on quadratic differential forms and dissipativity of linear differential systems

3 Dissipativity of linear differential systems

As was also noted in [9], the study of the algebraic Riccati equation can be put into the more general framework of studying quadratic storage functions for linear systems with quadratic supply rates. In that context, an important role is played by the *linear matrix inequality* $L(K) \geq 0$, where

$$L(K) := \begin{pmatrix} A^TK + KA + Q & KB + S^T \\ B^TK + S & R \end{pmatrix} \tag{44.6}$$

It can be shown that $L(K) \geq 0$ if and only if the quadratic function $V(x) := -x^TKx$ satisfies the inequality

$$\frac{d}{dt}V(x(t)) \leq x^T(t)Qx(t) + 2x^T(t)S^Tu(t) + u^T(t)Ru(t) \tag{44.7}$$

for all $t \in \mathbf{R}$, for all x and u satisfying the differential equation $\dot{x} = Ax+Bu$. The inequality (44.7) is called *the dissipation inequality* for the system $\dot{x} = Ax + Bu$ with *supply rate* $x^TQx + 2x^TS^Tu + u^TRu$. If a function $V(x)$ satisfies this inequality, it is called a *storage function.* The dissipation inequality expresses the property that along trajectories x and u of the system the increase in internal storage cannot exceed the rate at which storage is supplied to the system. If such a function $V(x)$ exists, we call the system *dissipative* with respect to the given supply rate. It can be shown that the system $\dot{x} = Ax + Bu$ with supply rate $x^TQx + 2x^TS^Tu + u^TRu$ is dissipative if and only if the frequency domain condition (44.3) (so along the imaginary axis) holds. Thus, (44.3) is equivalent to the existence of a real symmetric solution K to the linear matrix inequality $L(K) \geq 0$. Moreover, if this condition holds, there exist real symmetric solutions K^- and K^+ such that any real symmetric solution K satisfies $K^- \leq K \leq K^+$. The function $V_1(x) := -x^T K^+ x$ is then the *smallest*, and the function $V_2(x) := -x^T K^- x$ is the *largest* storage function. Also, K^- and K^+ are solutions of the algebraic Riccati equation (44.1). These considerations show that the existence of a negative semi-definite solution to (44.1) is equivalent to the existence of a negative semi-definite solution to the linear matrix inequality (44.6), equivalently, to the existence of a positive semi-definite storage function of the system $\dot{x} = Ax + Bu$ with supply rate $x^T Qx + 2x^T S^T u + u^T Ru$.

The general problem of the existence of storage functions was recently put in the framework of quadratic differential forms for linear differential systems [12], [8]). An important role in these references is played by *two-variable polynomial matrices*, i.e., matrices whose coefficients are real polynomials in two indeterminates, say ζ and η. A two-variable polynomial matrix Φ can be represented as

$$\Phi(\zeta, \eta) = \sum_{k,j} \Phi_{k,j} \zeta^k \eta^j$$

where the $\Phi_{k,j}$ are matrices with real coefficients, $k, j \in \mathbf{N}$, and the sum is a finite one. Φ is called symmetric if $\Phi(\zeta, \eta)^T = \Phi(\eta, \zeta)$. Each symmetric $q \times q$ two-variable poynomial matrix Φ induces a *quadratic differential form* (QDF), i.e., a map $Q_\Phi : \mathcal{C}^\infty(\mathbf{R}, \mathbf{R}^q) \to \mathcal{C}^\infty(\mathbf{R}, \mathbf{R})$, defined by

$$Q_\Phi(\ell) := \sum_{k,j} (\frac{d^k \ell}{dt^k})^T \Phi_{k,j} \frac{d^j \ell}{dt^j}$$

Associated with Φ, we define the (one-variable) polynomial matrix $\partial\Phi$ by $\partial\Phi(\xi) := \Phi(-\xi, \xi)$. Note that this polynomial matrix is para-hermitian, i.e., $(\partial\Phi(-\xi))^T = \partial\Phi(\xi)$. A QDF Q_Φ is called *average non-negative* if for all $\ell \in \mathcal{C}^\infty(\mathbf{R}, \mathbf{R}^q)$ of compact support we have

$$\int_{-\infty}^{\infty} Q_\Phi(\ell) dt \geq 0$$

In [12] this property was shown to be equivalent with the existence of a symmetric two-variable polynomial matrix Ψ such that

$$\frac{d}{dt} Q_\Psi(\ell) \leq Q_\Phi(\ell) \text{ for all } \ell \in \mathcal{C}^\infty(\mathbf{R}, \mathbf{R}^q) \qquad (44.8)$$

It was also shown that Q_Φ is average non-negative if and only if $\partial\Phi$ is non-negative along the imaginary axis, i.e., $\partial\Phi(i\omega) \geq 0$ for all $\omega \in \mathbf{R}$. We now explain how this result can be used to rederive the condition 44.3 for the existence of a real symmetric solution to the algebraic Riccati equation. Consider the controllable system $\dot{x} = Ax + Bu$, more precisely, the system $\Sigma = (\mathbf{R}, \mathbf{R}^{n+m}, \mathcal{B})$, with time axis $\mathbf{R}$, signal space $\mathbf{R}^{n+m}$ and behavior $\mathcal{B} := \{(x, u) \in \mathcal{C}^\infty(\mathbf{R}, \mathbf{R}^{n+m}) \mid \dot{x} = Ax + Bu\}$. Any controllable system also admits an image representation (see ([6, 11]). Consider the image representation

$$\begin{pmatrix} x \\ u \end{pmatrix} = \begin{pmatrix} C(\frac{d}{dt})B \\ p(\frac{d}{dt})I \end{pmatrix} \ell \tag{44.9}$$

Here, $p(\xi)$ is the characteristic polynomial of A, i.e., $p(\xi) := \det(\xi I - A)$, and $C(\xi)$ is the polynomial matrix defined by $C(\xi) := p(\xi)(\xi I - A)^{-1}$, that is, the classical adjoint of $\xi I - A$, appearing when one applies Cramer's rule to compute the inverse $(\xi I - A)^{-1}$. We have

$$\mathcal{B} = \{(x, u) \in \mathcal{C}^\infty(\mathbf{R}, \mathbf{R}^{n\times m}) \mid \text{ there exists } \ell \in \mathcal{C}^\infty(\mathbf{R}, \mathbf{R}^m) \text{ such that } 44.9\}$$

so (44.9) indeed defines an image representation of our system Σ. Now define the symmetric two-variable polynomial matrix Φ by

$$\Phi(\zeta, \eta) := \begin{pmatrix} C(\zeta)B \\ p(\zeta)I \end{pmatrix}^T \begin{pmatrix} Q & S^T \\ S & R \end{pmatrix} \begin{pmatrix} C(\eta)B \\ p(\eta)I \end{pmatrix} \tag{44.10}$$

It is immediate that if x, u and ℓ are related by (44.9), then for the QDF Q_Φ associated with Φ we have $Q_\Phi(\ell) = x^T Qx + 2x^T S^T u + u^T Ru$. Assume now that Q_Φ is average non-negative, equivalently, there exists Ψ such that $\frac{d}{dt}Q_\Psi \leq Q_\Phi$. It was proven in ([8]) that any such Q_Ψ can be represented as a (static) quadratic function of any state of the underlying system. In our case, a (minimal) state is given by $x = C(\frac{d}{dt})B\ell$, so there exists a real symmetric matrix $K \in \mathbf{R}^{n\times n}$ such that if x and ℓ are related by $x = C(\frac{d}{dt})B\ell$, then $Q_\Psi(\ell) = x^T Kx$. Collecting these facts we find that the following three statements are equivalent:

1. $\int_{-\infty}^{\infty}(x^T Qx + 2x^T S^T u + u^T Ru)dt \geq 0$ for all $x \in \mathcal{C}^\infty(\mathbf{R}, \mathbf{R}^n)$ and $u \in \mathcal{C}^\infty(\mathbf{R}, \mathbf{R}^m)$ of compact support, satisfying $\dot{x} = Ax + Bu$

2. there exists a real symmetric matrix $K \in \mathbf{R}^{n\times n}$ such that

$$\frac{d}{dt}x^T Kx \leq x^T Qx + 2x^T S^T u + u^T Ru$$

   for all $x \in \mathcal{C}^\infty(\mathbf{R}, \mathbf{R}^n)$ and $u \in \mathcal{C}^\infty(\mathbf{R}, \mathbf{R}^m)$ satisfying $\dot{x} = Ax + Bu$

3. $\partial\Phi(i\omega) \geq 0$ for all $\omega \in \mathbf{R}$

As noted before, condition (2) is equivalent to the existence of a real symmetric solution of the algebraic Riccati equation. Note that for Φ given by (44.10) we have $\Phi(\zeta, \eta) = p(\zeta)p(\eta)W(\zeta, \eta)$ with W the two-variable rational matrix given by (44.2). Hence, along the imaginary axis we have $\partial\Phi(i\omega) = |p(i\omega)|^2\partial W(i\omega)$. Therefore, condition (3) is equivalent to the frequency domain inequality (44.3). Thus we have re-established the fact that the existence of a real symmetric solution to the algebraic Riccati equation is equivalent to the frequency domain inequality (44.3).

4 The existence of negative semi-definite solutions

Let us now study what the set-up of QDF's for linear differential systems can tell us on the existence of negative semi-definite solutions to the algebraic Riccati equation. To start with, let Φ be an arbitrary $q \times q$ symmetric two-variable polynomial matrix. The associated QDF Q_Φ is called *half-line non-negative* if for all $\ell \in \mathcal{C}^\infty(\mathbf{R}, \mathbf{R}^q)$ of compact support we have

$$\int_{-\infty}^{0} Q_\Phi(\ell)dt \geq 0$$

It was proven in [12] that a QDF Q_Φ is half-line non-negative if and only if there exists a symmetric two-variable polynomial matric Ψ such that $Q_\Psi \geq 0$ and such that (44.8) holds. Applying this fact to our system Σ, again using that it has an image representation given by (44.9), we this time find that the following two statements are equivalent:

1. $\int_{-\infty}^{0}(x^TQx + 2x^TS^Tu + u^TRu)dt \geq 0$ for all $x \in \mathcal{C}^\infty(\mathbf{R}, \mathbf{R}^n)$ and $u \in \mathcal{C}^\infty(\mathbf{R}, \mathbf{R}^m)$ of compact support, satisfying $\dot{x} = Ax + Bu$

2. there exists a real symmetric matrix $K \in \mathbf{R}^{n\times n}$, $K \leq 0$, such that $\frac{d}{dt}(-x^TKx) \leq x^TQx + 2x^TS^Tu + u^TRu$ for all $x \in \mathcal{C}^\infty(\mathbf{R}, \mathbf{R}^n)$ and $u \in \mathcal{C}^\infty(\mathbf{R}, \mathbf{R}^m)$ satisfying $\dot{x} = Ax + Bu$

As before, condition (2) is equivalent to the existence of a negative semi-definite real symmetric solution to the algebraic Riccati equation. In the following, we establish a condition on the two-variable polynomial matrix Φ that is equivalent to condition (1). It turns out that in this way we reobtain the frequency domain condition for the existence of a negative semi-definite solution to the algebraic Riccati equation that was obtained before in [4]. First note that condition (1) is equivalent to the condition that for all $\ell \in \mathcal{C}^\infty(\mathbf{R}, \mathbf{R}^m)$ of compact support we have

$$\int_{-\infty}^{0} \begin{pmatrix} A(\frac{d}{dt})B\ell \\ p(\frac{d}{dt})\ell \end{pmatrix}^T \begin{pmatrix} Q & S^T \\ S & R \end{pmatrix} \begin{pmatrix} A(\frac{d}{dt})B\ell \\ p(\frac{d}{dt})\ell \end{pmatrix} dt \geq 0 \qquad (44.11)$$

By an approximation argument, this condition is equivalent to the same inequality holding for all $\ell \in \mathcal{C}^\infty(\mathbf{R}, \mathbf{R}^m)$ such that $\int_{-\infty}^{0} \|\ell\|^2 dt$ is finite. Now, let $N \in \mathbf{N}$, and let $\lambda_1, \lambda_2, \ldots, \lambda_N$ be N distinct complex numbers in $\Re e(\lambda) > 0$. Also, let $v_1, v_2, \ldots v_N$ be arbitrary vectors in $\mathbf{C}^m$. Consider the function

$$\ell(t) := \sum_{i=1}^{N} e^{\lambda_i t} v_i \qquad (44.12)$$

Applying (44.11) to this function ℓ (silently moving from real valued functions to complex valued functions), we obtain

$$\sum_{i=1}^{N}\sum_{j=1}^{N}\int_{-\infty}^{0}(e^{(\bar{\lambda_i}+\lambda_j)t}v_i^*\begin{pmatrix} A(\bar{\lambda_i})B \\ p(\bar{\lambda_i})I \end{pmatrix}^T\begin{pmatrix} Q & S^T \\ S & R \end{pmatrix}\begin{pmatrix} A(\lambda_j)B \\ p(\lambda_j)I \end{pmatrix}v_j)dt \geq 0$$

which, after integration, yields

$$\sum_{i=1}^{N}\sum_{j=1}^{N} v_i^*\frac{\Phi(\bar{\lambda_i},\lambda_j)}{\bar{\lambda_i}+\lambda_j}v_j \geq 0 \tag{44.13}$$

Since, for fixed complex numbers $\lambda_1, \lambda_2, \ldots, \lambda_N$, this holds for *all* $v_1, v_2, \ldots, v_N$, this implies that the hermitian $mN \times mN$ matrix whose (i, j)th block is equal to the $m \times m$ matrix $\frac{\Phi(\bar{\lambda_i},\lambda_j)}{\bar{\lambda_i}+\lambda_j}$ is positive semi-definite. If, conversely, (44.13) holds for any choice of N, any choice of distinct complex numbers $\lambda_1, \lambda_2, \ldots, \lambda_N$ in $\Re e(\lambda) > 0$, and any choice of vectors $v_1, v_2, \ldots v_N$ in $\mathbf{C}^m$, then clearly (44.11) holds for any function ℓ of the form (44.12). Again by an approximation argument, the inequality (44.11) must then hold for all $\ell \in \mathcal{C}^\infty(\mathbf{R}, \mathbf{R}^m)$ of compact support. This proves that any of the statements (1) and (2) is equivalent with the following condition (3):

3. for all $N \in \mathbf{N}$, and for any choice $\lambda_1, \lambda_2, \ldots, \lambda_N$ of distinct complex numbers in $\Re e(\lambda) > 0$, the hermitian $mN \times mN$ matrix

$$\left(\frac{\Phi(\bar{\lambda_i},\lambda_j)}{\bar{\lambda_i}+\lambda_j}\right)_{i,j=1,2,\ldots N}$$

   is positive semi-definite.

We now express condition (3) in terms of the two-variable rational matrix W. It is easily verified that

$$\left(\frac{\Phi(\bar{\lambda_i},\lambda_j)}{\bar{\lambda_i}+\lambda_j}\right)_{i,j=1,2,\ldots,N} = D^*\left(\frac{W(\bar{\lambda_i},\lambda_j)}{\bar{\lambda_i}+\lambda_j}\right)_{i,j=1,2,\ldots,N} D$$

Here D is the blockdiagonal matrix whose ith diagonal block is equal to $p(\lambda_i)I$ (with I the $m \times m$ identity matrix). Thus we claim that condition (3) is equivalent to the following condition (4):

4. for all $N \in \mathbf{N}$, and for any choice $\lambda_1, \lambda_2, \ldots, \lambda_N$ of distinct complex numbers in $\Re e(\lambda) > 0$ such that $\lambda_i \notin \sigma(A)$, $i = 1, 2, \ldots, N$, the hermitian $mN \times mN$ matrix $\left(\frac{W(\bar{\lambda_i},\lambda_j)}{\bar{\lambda_i}+\lambda_j}\right)_{i,j=1,2,\ldots N}$ is positive semi-definite.

Indeed, if $\lambda_i \notin \sigma(A)$, $i = 1, 2, \ldots, N$, then D is nonsingular. From this the proof of the implication (3) $\Rightarrow$ (4) is obvious. The proof of the converse implication then follows by using a continuity argument.

We have now shown that the existence of a negative semi-definite real symmetric solution of the algebraic Riccati equation is equivalent to condition (4) on the two-variable rational matrix W. Thus we have re-established the condition that was proposed in [4] as a correct alternative for the errorous frequency domain condition (44.4). A similar condition was obtained in [1] in related work on spectral factorization.
We remind the reader that the title of this note is: "When does the algebraic Riccati equation have a negative semi-definite solution?" We have re-derived the necessary and sufficient frequency domain condition (4) for this to hold. Yet, in our opinion, this condition is not a satisfactory one, since not only it requires one to check non-negativity of an infinite number of hermitian matrices, but also there is no upper bound to the dimension of these matrices. In view of this, we formulate the following open problem:

> **Find a reasonable necessary and sufficient frequency domain condition, i.e., a condition in terms of the rational matrix ∂W, or possibly in terms of the two-variable rational matrix W, for the existence of a real symmetric negative semi-definite solution of the algebraic Riccati equation (44.1).**

5 References

[1] B.D.O. Anderson, "Corrections to: Algebraic properties of minimal degree spectral factor". *Automatica*, Vol. 11, pp. 321 - 322, 1975.

[2] B.D.O. Anderson and S. Vongpanitlerd, *Network Analysis and Synthesis - A Modern Systems Theory Approach*, Prentice-Hall, Englewood Cliffs, N.J., 1973.

[3] S. Bittanti, A.J. Laub, J.C. Willems (Eds.), *The Riccati Equation*, Springer Verlag, Berlin, 1991.

[4] B.P. Molinari, "Conditions for non-positive solutions of the linear matrix inequality". *IEEE Trans. Automat. Contr.*, Vol. AC-29, pp. 804 - 806, 1975.

[5] P.J. Moylan, "On a frequency condition in linear optimal control theory". *IEEE Trans. Automat. Contr.*, Vol. AC-29, pp. 806, 1975.

[6] J.W. Polderman and J.C. Willems, *Introduction to Mathematical Systems Theory*, Springer Verlag, 1997.

[7] V.M. Popov, *L'Hyperstabilité des systèmes automatiques*, Dunod, Paris, 1973.

[8] H.L. Trentelman and J.C. Willems, "Every storage function is a state function", *Systems and Control Letters*, 32, pp. 249 - 259, 1997.

[9] J.C. Willems, "Least squares stationary optimal control and the algebraic Riccati equation", *IEEE Trans. Automat. Contr.*, Vol. AC-16, pp. 621 - 634, 1971.

[10] J.C. Willems, "On the existence of a nonpositive solution to the Riccati equation", *IEEE Trans. Automat. Contr.*, Vol. AC-19, pp. 592 - 593, 1974

[11] J.C. Willems, "Paradigms and puzzles in the theory of dynamical systems", *IEEE Trans. Automat. Contr.*, Vol. 36, pp. 259-294, 1991.

[12] J.C. Willems and H.L. Trentelman, "On quadratic differential forms". To appear in *SIAM J. Contr. Optim.*.

45

Representing a nonlinear input-output differential equation as an input-state-output system

A.J. van der Schaft

University of Twente
Faculty of Applied Mathematics
P.O. Box 217
7500 AE Enschede
THE NETHERLANDS
a.j.vanderschaft@math.utwente.nl

1 Introduction

It is well-known that every *linear* system of higher-order differential equations

$$P_0 y + P_1 \dot{y} + \ldots + P_k y^{(k)} = Q_0 u + Q_1 \dot{u} + \ldots + Q_k u^{(k)} \tag{45.1}$$

with $y \in \mathbb{R}^p, u \in \mathbb{R}^m$, and $P(s) := P_0 + P_1 s + \ldots + P_k s^k$ a $p \times p$ polynomial matrix with $\det P(s) \not\equiv 0$, can be represented (*realized*) by a minimal state space system

$$\begin{array}{rcll} \dot{x} & = & Ax + Bu & , x \in \mathbb{R}^n \\ y & = & Cx + Du & \end{array} \tag{45.2}$$

The corresponding question for a *nonlinear* system of higher-order differential equations

$$F(y, \dot{y}, \ldots, y^{(k)}, u, \dot{u}, \ldots, u^{(k)}) = 0 \tag{45.3}$$

with $y \in \mathbb{R}^p, u \in \mathbb{R}^m$, and F mapping into $\mathbb{R}^p$, is much harder. In fact, (45.3) can be represented as a nonlinear state space system

$$\begin{array}{rcll} \dot{x} & = & f(x,u), & x \in \mathcal{X} \quad n-\text{dimensional manifold} \\ y & = & h(x,u) & \end{array} \tag{45.4}$$

only under severe (non-generic) "integrability" conditions on the mapping F. This was shown for $k = 1$ in [1], and extended to general k in [2], see also

[3]. Some feeling for the required integrability conditions can be developed by noting that in eliminating the state variables x from (45.4) one will successively *differentiate* the output equations to

$$\dot{y} = \frac{\partial h}{\partial x}(x,u)f(x,u) + \frac{\partial h}{\partial u}(x,u)\dot{u}, \quad \ddot{y} = \ldots,$$

etc., and then eliminate x from these equations. Since these differentiated output equations have a special structure, one will thus obtain equations (45.3) which are also of a special structure. This special structure is captured in the integrability conditions derived in [2]. For connections with state space representations of *input-output maps* we refer e.g. to [4].
Although the conditions for representability of (45.3) as an input-state-output system (45.4) as derived in [2] are sufficient and necessary (modulo constant rank assumptions), they are not very transparant. Moreover the derived conditions are *algorithm-oriented*, in the sense that at every step of the developed state space representation algorithm necessary and sufficient integrability conditions have to be checked. Thus there is a clear need for a more abstract approach. The present note points to such an approach.

2 The linear case

Let us first consider the *linear* situation. For compactness of notation, write $w = \begin{bmatrix} y \\ u \end{bmatrix} \in \mathbb{R}^q, q = m + p$, and denote (45.1) by

$$R\left(\frac{d}{dt}\right) w(t) = 0 \tag{45.5}$$

with $R(s) = [P_0 \vdots - Q_0] + [P_1 \vdots - Q_1]s + \ldots + [P_k \vdots - Q_k]s^k$.
Question: what is a minimal state space of (45.5)? Denote the set of weak solutions of (45.5) by $\mathcal{B}$ (see e.g. [6], [5] for the terminology). Let $w \in \mathcal{B}$. Define $\overline{w} = w \wedge 0$ as the *concatenation* of w with the zero trajectory in $\mathbb{R}^q$, i.e. $\overline{w}(t) = w(t), t < 0$, and $\overline{w}(t) = 0, t \geq 0$. Then $\overline{w}$ is again a weak solution of (45.5) (corresponding to the *zero* state at $t = 0$) if and only if

$$\int_{-\infty}^{+\infty} \overline{w}^T(t) R^T(-\frac{d}{dt})\varphi(t)dt = 0 \tag{45.6}$$

for all C^∞ (test) functions φ with compact support. Take for simplicity w to be C^∞. Then by partial integration

$$\begin{aligned} &\textstyle\int_{-\infty}^{+\infty} \overline{w}^T(t) R^T(-\frac{d}{dt})\varphi(t)dt = \int_{-\infty}^{0} w^T(t) R^T(-\frac{d}{dt})\varphi(t)dt = \\ &\textstyle\int_{-\infty}^{0} \varphi^T(t) R(\frac{d}{dt}) w(t)dt + \text{ remainder at } 0 = \text{remainder at } 0 \end{aligned} \tag{45.7}$$

since $R(\frac{d}{dt})w(t) = 0$. Furthermore, the remainder has the form

$$\left[\varphi^T(t)\dot{\varphi}^T(t)\ldots\varphi^{(k-1)^T}(t)\right] Q \begin{bmatrix} w(t) \\ \dot{w}(t) \\ \vdots \\ w^{(k)}(t) \end{bmatrix} \tag{45.8}$$

for some constant matrix Q of appropriate dimensions (in fact, the last q columns of Q will be zero). We conclude that (45.6) holds for all test functions φ if and only if

$$\begin{bmatrix} w(0) \\ \dot{w}(0) \\ \vdots \\ w^{(k)}(0) \end{bmatrix} \in \ker Q \tag{45.9}$$

By the general theory of Willems (see e.g. [5]) the minimal state space $\mathcal{X}$ is given as $\mathcal{B}/\mathcal{B}_0$, where $\mathcal{B}_0$ is the set of $w \in \mathcal{B}$ such that $w \wedge 0$ is in $\mathcal{B}$. Denote now by Z be the linear space of points $(w, \dot{w}, \ldots, w^{(k)}) \in \mathbb{R}^{(k+1)q}$ satisfying (45.5), that is

$$R_0 w + R_1 \dot{w} + \ldots + R_k w^{(k)} = 0$$

with $R_i = [P_i \vdots -Q_i]$. Then it follows from the above that the minimal state space $\mathcal{X}$ is also given as

$$\mathcal{X} = Z / \ker Q \cap Z \tag{45.10}$$

This result can be seen to be equivalent to the one obtained in [6].

3 The nonlinear case

Let us now try to extend this idea to the nonlinear case. We start from the partial integration (45.7). However, as suggested e.g. in [7] we first need to consider the *variational* (or linearized) *systems* corresponding to (45.3), that is

$$\frac{\partial F}{\partial y} y_v + \frac{\partial F}{\partial \dot{y}} \dot{y}_v + \ldots + \frac{\partial F}{\partial u} u_v + \frac{\partial F}{\partial \dot{u}} \dot{u}_v + \ldots + \frac{\partial F}{\partial u^{(k)}} u_v^{(k)} = 0 \tag{45.11}$$

where $u_v \in \mathbb{R}^m$ and $y_v \in \mathbb{R}^p$ are the *variational* inputs and outputs. Here, of course, the partial Jacobians $\frac{\partial F}{\partial y}, \frac{\partial F}{\partial \dot{y}}, \ldots, \frac{\partial F}{\partial u}, \ldots$ etc., still depend on the variables $(y, \dot{y}, \ldots, y^{(k)}, u, \dot{u}, \ldots, u^{(k)})$.

Repeated partial integration of these variational systems multiplied by a test vector φ yields (see [7])

$$\int_{t_1}^{t_2} \varphi^T(t)(\tfrac{\partial F}{\partial y} y_v + \ldots + \tfrac{\partial F}{\partial u^{(k)}} u_v^{(k)})(t)dt =$$

$$\int_{t_1}^{t_2} [y_v^T(t) S_1(\varphi, \dot{\varphi}, \ldots, \varphi^{(k)})(t) + u_v^T(t) S_2(\varphi, \dot{\varphi}, \ldots, \varphi^{(k)})(t)]dt$$

$$+ [\varphi^T(t) \ \dot{\varphi}^T(t) \ldots \varphi^{(k-1)^T}(t)] Q(y, \ldots, y^{(k)}, u, \ldots, w^{(k)}) \begin{bmatrix} w_v(t) \\ \dot{w}_v(t) \\ \vdots \\ w_v^{(k)}(t) \end{bmatrix}]_{t_1}^{t_2} \tag{45.12}$$

where $w_v(t) = \begin{bmatrix} y_v(t) \\ u_v(t) \end{bmatrix}$, for a certain matrix $Q(y, \ldots, y^{(k)}, u, \ldots, u^{(k)})$.
Here the terms S_1 and S_2 determine the *adjoint* variational systems [7], but this is not our present interest.
As in the linear case we can consider the *kernel* of the matrix Q; however in the nonlinear case this kernel space will depend on

$$(y, \dot{y}, \ldots, y^{(k)}, u, \dot{u}, \ldots, u^{(k)}),$$

since Q does. A proper formulation of this is to regard

$$(y, \dot{y}, \ldots, y^{(k)}, u, \dot{u}, \ldots, u^{(k)})$$

as points in the k-jet space $T_k(Y \times U)$ (which, since $Y = \mathbb{R}^p, U = \mathbb{R}^m$, is isomorphic with $\mathbb{R}^{(k+1)q}, q = p + m$), and to define a *distribution* D on $T_k(Y \times U)$ by letting

$$D(y, \dot{y}, \ldots, y^{(k)}, u, \dot{u}, \ldots, u^{(k)}) := \ker Q(y, \ldots, y^{(k)}, u, \ldots, u^{(k)}) \tag{45.13}$$

The main open problem suggested in this note is to prove or disprove the following

Conjecture 17 *Assume the distribution $D \cap TZ$ on the submanifold*

$$Z := \{(y, \dot{y}, \ldots, y^{(k)}, u, \dot{u}, \ldots, u^{(k)}) \in T_k(Y \times U) |$$
$$F(y, \dot{y}, \ldots, y^{(k)}, u, \dot{u}, \ldots, u^{(k)}) = 0\}$$

, with D defined by (45.13), is constant-dimensional. Assume for simplicity that the Jacobian $\frac{\partial F}{\partial y^{(k)}}$ has full rank. Then the system (45.3) admits a state space representation (45.4) if and only if the distribution $D \cap TZ$ is involutive on Z. Furthermore, a minimal state space $\mathcal{X}$ is given as Z factored out by the leaves of this involutive distribution.

Remark 18 *Obviously the condition that rank* $\frac{\partial F}{\partial y^{(k)}} = p$ *can be relaxed. Furthermore, a similar conjecture can be stated if we do not a priori distinguish the external variables into inputs u and outputs y.*

The conjecture is true in the following case already treated in [1]. Consider the equation

$$\dot{y} - a(y, u, \dot{u}) = 0 \quad , \quad u, y \in \mathbb{R} \tag{45.14}$$

In this case, the remainder is given as $\varphi y_v - \varphi \frac{\partial a}{\partial \dot{u}} u_v$. Thus the distribution D on $T_1(\mathbb{R} \times \mathbb{R})$ is given as

$$D = \text{span}\,\{\frac{\partial}{\partial \dot{y}}, \frac{\partial}{\partial \ddot{u}}, \frac{\partial a}{\partial \dot{u}}\frac{\partial}{\partial y} + \frac{\partial}{\partial u}\}$$

It follows directly from the results of [1], [2] that (45.14) can be represented by an input-state-output system (45.4) if and only if $D \cap TZ$ is involutive.

4 References

[1] M.I. Freedman, J.C. Willems, "Smooth representations of systems with differentiated inputs", *IEEE Trans. Aut. Contr. 23 (1978), 16-22*

[2] A.J. van der Schaft, "On realization of nonlinear systems described by higher-order differential equations", *Math. Syst. Theory, 19 (1987), 239-275*

[3] E. Delaleau, W. Respondek, "Lowering the orders of derivatives of controls in generalized state space systems", *J. Math. Systems Estim. Contr. 5 (1995), 1-27*

[4] Y. Wang, E. Sontag, "Generating series and nonlinear systems, analytic aspects, local realizability, and i/o representations", *orum Mathematicum 4 (1992), 299-322*

[5] J.C. Willems, "Models for dynamics", *Dynamics Reported, 2 (1989), 171-269*

[6] P. Rapisarda, J.C. Willems, "State maps for linear systems", *SIAM J. Contr. Opt., 35 (1997), 1053-1091*

[7] P.E. Crouch, F. Lamnabhi-Lagarrigue, A.J. van der Schaft, "Adjoint and Hamiltonian input-output differential equations", *IEEE Trans. Aut. Contr. 40 (1995), 603-615*

46

Shift policies in QR-like algorithms and feedback control of self-similar flows

Paul Van Dooren* and Rodolphe Sepulchre**

*Université catholique de Louvain
Av. Lemaitre, 4
1348 Louvain-la-Neuve
BELGIUM
vdooren@csam.ucl.ac.be

**Institut Montefiore
University of Liège B28
B-4000 Liège
BELGIUM
sepulchre@montefiore.ulg.ac.be

1 Description of the problem

***FG* algorithms.** *FG* algorithms are generalizations of the well-known *QR* algorithm for calculating the eigenvalues of a matrix. Let $\mathcal{F}$ and $\mathcal{G}$ two closed subgroups of the general linear group $GL_n(\mathbb{F})$ ($\mathbb{F} = \mathbb{R}$ or $\mathbb{C}$). Assuming $\mathcal{F} \cap \mathcal{G} = \{I\}$, each matrix $A \in GL_n(\mathbb{F}))$ has at most one factorization of the form $A = FG$, where $F \in \mathcal{F}$ and $G \in \mathcal{G}$. Starting from a given matrix $B_0 \in GL_n(\mathbb{F}))$, the *FG algorithm* produces a sequence of matrices B_m, $m = 1, 2, \ldots$, as follows: B_i is factored into a product $B_i = F_{i+1}G_{i+1}$ and this product is reversed to define $B_{i+1} := G_{i+1}F_{i+1}$. Thus

$$B_i = F_{i+1}G_{i+1} \Rightarrow B_{i+1} := G_{i+1}F_{i+1}. \tag{46.1}$$

What is typically expected from the *FG* algorithm is that the sequence of iterates $\{B_i\}_{i \geq 1}$ converges to a matrix whose diagonal entries are the eigenvalues of B_0. The sequence is indeed a sequence of similarity transformations, that is

$$B_{i+1} = F_{i+1}^{-1}B_iF_{i+1} = G_{i+1}B_iG_{i+1}^{-1}. \tag{46.2}$$

A key property of the algorithm is

$$B_m = F_{(m)}^{-1} B_0 F_{(m)} = G_{(m)} B_0 G_{(m)}^{-1} \quad (46.3)$$

$$B_0^m = F_{(m)} G_{(m)} \quad (46.4)$$

where the notation $F_{(m)}$ stands for the product $F_m F_{m-1} \dots F_1$.

Shifted FG algorithms. In general, some iterate B_i of the algorithm may fail to have an FG factorization, in which case the algorithm breaks down. In addition, even if the algorithm converges, the convergence may be very slow. These two reasons motivate the introduction of *shift policies* in the standard FG algorithm. The simplest modification is of the form

$$B_i - \sigma_i I = F_{i+1} G_{i+1} \Rightarrow B_{i+1} := G_{i+1} F_{i+1} + \sigma_i I$$

which still ensures the self-similarity $B_{i+1} = F_{i+1}^{-1} B_i F_{i+1} = G_{i+1} B_i G_{i+1}^{-1}$. More general shift policies allow a polynomial

$$p_i(B_i) = \prod_j (B_i - \sigma_{i_j} I)$$

to replace B_i in the standard algorithm. This allows for several steps of the classic FG algorithm to be concatenated in one step of the generalized FG algorithm.

Shifted FG algorithms typically lead to improved convergence in the situations where the FG algorithm converges. In the situations where the FG algorithm does not converge, shift policies may be used to achieve a continuation of the algorithm, by avoiding certain singularities ([6]).

Open problem. The *analysis* of shifted algorithms and a systematic *design* of shift policies to improve convergence of the FG algorithms are essentially open problems in numerical analysis (See Section 4 for some existing results).

2 Reformulation of the problem

Self-similar flows. Self-similar flows are the solutions of differential equations that can be associated with FG algorithms because of the Lie group structure of the closed subgroups of $GL_n(\mathbb{F})$. Under (weak) extra conditions, self-similar flows continuously interpolate the solution of the FG algorithm, that is, the self-similar flow coincides at integer times with the iterates of the FG algorithm.

Let $\Lambda(\mathcal{F})$ and $\Lambda(\mathcal{G})$ be the Lie algebras associated with the Lie groups $\mathcal{F}$ and $\mathcal{G}$. Suppose that $\mathbb{F}^{n\times n} = \Lambda(\mathcal{F}) \oplus \Lambda(\mathcal{G})$. Then every matrix $M \in \mathbb{F}^{n\times n}$ can be expressed uniquely as a sum

$$M = \rho(M) + \nu(M)$$

where $\rho(M) \in \Lambda(\mathcal{F})$ and $\nu(M) \in \Lambda(\mathcal{G})$. Let $f(.)$ a function defined on $sp(B)$ where $sp(B)$ denotes the spectrum of B (this means $f(.)$ and its n first derivatives must be defined on an open set containing $sp(B)$). Then the solution of the differential equation

$$\dot{B} = [B, \rho(f(B))] = [\nu(f(B)), B], \; B(0) = B_0 \tag{46.5}$$

is self-similar, i.e. it satisfies

$$B(t) = F(t)^{-1} B_0 F(t) = G(t) B_0 G(t)^{-1}, \tag{46.6}$$

where F and G are solutions of

$$\dot{F} = F\rho(f(B(t)), \; F(0) = I$$

and

$$\dot{G} = G\nu(f(B(t)), \; G(0) = I.$$

The key property satisfied by these differential equations is

$$\exp(f(B_0)t) = F(t)G(t). \tag{46.7}$$

Property (46.6) is the continuous analogue of (46.3) while (46.7) is the continuous analogue of (46.4). The interpolating property of self-similar flows is obtained by choosing $f(\cdot) = \log(\cdot)$.
In shifted FG algorithms, the property (46.4) is replaced by a more general equation

$$p_m(B_0) \ldots p_2(B_0) p_1(B_0) = F_{(m)} G_{(m)},$$

where the successive polynomials p_i are determined by the shift policy. The continuous analogue is obtained by switching at integer times between different functions $f_i(\cdot) = \log(p_i(\cdot))$. The shift policy is typically a feedback process since the shift is defined at iteration i as a function of the current "state" B_i. As a consequence, a shift policy defines a feedback control strategy for the nonlinear differential equation (46.5), the control variable being the function f. If the control f is switched at integer times based on some switching logic, a hybrid feedback control system is associated to the shifted FG algorithm. This reformulation leads to open questions of the following type:

- (Analysis) What are the limit sets and the convergence and robustness properties of the differential equation (46.5) ?

- (Synthesis) Design a switching controller which guarantees the absence of finite escape time (continuous analogue of the FG algorithm breaking down after a finite number of iterations) and ensures convergence to a desired equilibrium (stabilization problem).

3 Motivation and history of the problem

The existence of continuous analogues of matrix algorithms was noticed in early developments of numerical analysis (Rutishauser himself [5] derived a continuous analogue of the quotient-difference algorithm, a predecessor of the QR algorithm). The complete connection between (shifted) FG algorithms and their corresponding self-similar flows is more recent and is due to Watkins and Elsner [6].
The potential importance of such connections was brought to the attention of the control community by Brockett [2, 3] who established the connection between the "double-bracket" flow and several algorithmic problems, including the QR algorithm. This line of research has quickly developed over the last few years and is reported in a recent book by Helmke and Moore [4]. The double-bracket flow reveals the gradient nature of the self-similar flow associated with the QR algorithm (for symmetric matrices with distinct eigenvalues), thereby connecting FG algorithms to other optimization problems defined on Riemannian manifolds.

4 Available results and desired extensions

The QR like algorithms always start with a preliminary reduction to a matrix $B_1 = F_1^{-1}B_0F_1 = G_1B_0G_1^{-1}$ in so-called condensed form. For the QR algorithm his e.g. a Hessenberg form for general matrices and a tridiagonal form for symmetric or Hermitian matrices. These condensed forms are very useful since they allow to *estimate* a particular eigenvalue, which is then used as shift. In the case of real matrices one uses typically second order polynomial with two complex conjugate shifts. This procedure works very well for tridiagonal matrices where appropriate shift techniques have been proved to yield global and/or cubic convergence [7]. For Hessenberg matrices the results are already less complete and for the more general class of FG algorithms there are no solid results available : for some shifts, the algorithm may even break down because the factorization (46.1) may not exist.
So far, the connections between matrix algorithms and their continuous-time analogues were always made a posteriori and did not lead to results of direct use for numerical analysts. The situation may be different in the study of shift policies. The geometric study of the double bracket flow [1, 3] indicates that the current developments of modern mechanics may help understanding the dynamics of self-similar flows. In addition, the control of hybrid systems is a quickly expanding research area including the hybrid control of continuous-time systems. For these reasons, the authors have some hope that the study of shift policies from a control theoretic perspective will contribute to the field of numerical analysis.

Acknowledgements

During the reviewing process of this manuscript, we were pointed out that Uwe Helmke (Würzburg University) had presented similar ideas at the MTNS conference of 1996 in St. Louis, but these were not reported in conference proceedings. He mentioned the issue of controllability of shifted QR algorithms, which we believe to be an important one.
This paper presents research results of the Belgian Programme on Interuniversity Poles of Attraction, initiated by the Belgian State, Prime Minister's Office for Science, Technology and Culture. The scientific responsibility rests with its authors.

5 References

[1] A.M. Bloch, R.W. Brockett, and T. Ratiu, "Completely Integrable Gradient Flows", *Communications in Mathematical Physics*, 147, pp. 57-74 (1992).

[2] R.W. Brockett, "Dynamical Systems That Sort Lists, Diagonalize Matrices and Solve Linear Programming Problems", *Linear Algebra and its Applications*, 146, pp. 79-91 (1991).

[3] R.W. Brockett, "Differential geometry and the design of gradient algorithms", *Proc. Symp. Pure Math.*, AMS 54, part I, pp. 69-92.

[4] U. Helmke and J. B. Moore, *Optimization and Dynamical Systems*, Springer-Verlag, London, 1994.

[5] H. Rutishauser, "Ein infinitesimales Analogon zum Quotienten-Differenzen-Algorithmus", *Arch. Math.* 5, pp. 132-137 (1954).

[6] D. Watkins and L. Elsner, "Self-Similar Flows", *Linear Algebra and its Applications*, 110, pp. 213-242 (1988).

[7] J. H. Wilkinson, *The Algebraic Eiegenvalue Problem*, Oxford Univ. Press, Oxford, 1965.

47

Equivalences of discrete-event systems and of hybrid systems

J.H. van Schuppen

CWI
P.O. Box 94079
1090 GB Amsterdam
THE NETHERLANDS
J.H.van.Schuppen@cwi.nl

1 Description of the problem

The problem is to develop concepts and theorems for equivalences of discrete-event systems and of hybrid systems.
For the equivalence of a discrete-event system consider a partially-observed deterministic automaton

$$A_1 = (Q_1, \Sigma_1, \delta_1, q_{10}, Q_{1m}, \Sigma_{1o}, p_1),$$

where Q_1 is an infinite discrete set called the state set; Σ_1 is a finite set called the event set; $\delta_1 : Q_1 \times \Sigma_1 \to Q_1$ is a partial function called the transition function; $q_{10} \in Q_1$ is the initial state; $Q_{1m} \subseteq Q_1$ is a subset of the state set representing accept states; Σ_{1o} is a finite set of observable events; Σ_1^* is the set of all finite strings with elements in Σ_1 and the empty string ϵ; and $p_1 : \Sigma_1^* \to \Sigma_{1o}^*$ is a causal map representing the observation map of the system.
The observation map is discussed in detail. Let Σ and Σ_0 be two alphabets and $L \subset \Sigma^*$ be a language. The map $p : L \to \Sigma_o^*$ is said to be causal if $p(\epsilon) = \epsilon$ and for all $s \in L$ and for all $\sigma \in \Sigma$ such that $s\sigma \in L$, either $p(s\sigma) = p(s)$ or there exists a $\sigma_o \in \Sigma_o$ such that $p(s\sigma) = p(s)\sigma_o \in \Sigma_o^*$. A projection is a special case of a causal map. The map $p : \Sigma^* \to \Sigma_o^*$, where $\Sigma_o \subseteq \Sigma$, is said to be projection if $p(\epsilon) = \epsilon$ and for $s \in \Sigma^*$, $p(p(s)) = p(s)$. A causal map is equivalent to a prefix-preserving map. See for references on causal and prefix-preserving maps [31] and [29, p. 248].
It is essential to the problem that the state set Q_1 is infinite and that the map p_1 is not injective.

Define the languages

$$\begin{aligned} L(A_1) &= \{s \in \Sigma_1^* \mid \delta_1(q_{10}, s) \in Q_{1m}\}, \\ p_1(L(A_1)) &= \{r \in \Sigma_{1o}^* \mid \exists\ s \in L^*(A_1) \text{ such that } r = p_1(s)\}. \end{aligned}$$

The problem is to formulate conditions on the discrete-event system A_1 such that there exists a partially-observed deterministic automaton

$$A_2 = (Q_2, \Sigma_2, \delta_2, q_{20}, Q_{2m}, \Sigma_{1o}, p_2),$$

with a *finite* state set Q_2 such that

$$p_1(L(A_1)) = p_2(L(A_2)). \tag{47.1}$$

It will be nice if the conditions have a system theoretic interpretation. Questions of the problem include:

1. What are necessary and sufficient conditions for the existence of the second system?

2. For which other concepts of equivalence than language equivalence as in (47.1) does the problem admit a solution that is moreover useful for problems of control and system theory?

3. How to exploit the equivalences and the corresponding classes for problems of control and system theory of discrete-event systems and of hybrid systems?

The emphasis in the problem is on the formulation of the concepts of equivalence, on system theoretic concepts for the problem, and on the use of equivalences for control and system theoretic problems. The emphasis is not exclusively on the computational properties of the solution. For particular instances the problem or the conditions of its solution are likely to be undecidable.

If one allows the state of the second system to be directly observable (p_2 is the identity map) then it is known that the equivalence exists iff the language $p_1(L(A_1))$ is regular, see [19, Th. 1.28]. Note however that the conditions asked for are to be formulated in terms of the system A_1 because in control theory that is the way the problem arises.

A hint for the problem is to think of decompositions of the state set as in decompositions of automata and of linear multivariable systems. The concept of decomposition of an automaton has been developed by K. Krohn and J. Rhodes, see [13]. For an exposition on decompositions see [10].

General references include: control and system theory [22]; automata and languages [10, 17, 19, 26, 30]; and control of discrete-event systems [16, 25, 29].

2 Motivation and history of the problem

The problem is motivated by problems of control and system theory for discrete-event systems and for hybrid systems. The aim of system theory is to identify classes of dynamic systems that admit the development of a theoretically rich and practically useful theory. Because a discrete-event system does have neither an analytic nor an algebraic structure, the currently existing system theory for finite-dimensional linear systems is not always an appropriate source of inspiration. System theorists may take inspiration from theoretical computer science but will have to adjust the available theory to control and system theoretic modeling.

Problems of signal representation, of the construction of observers, or of control synthesis can with some creativity be reformulated as realization problems, including the construction of equivalent systems. In addition, in control of discrete-event systems one considers the relation between the closed-loop system and the control objectives. The control objectives are often presented in the form of a language.

Conditions may have to be imposed on the second system to make the problem useful for engineering problems. Decidability of problems and limitation of the complexity of algorithims is of major importance to control and system theory. In the theory of computation a language is called *decidable* if there exists a Turing machine such that when any string of that language is fed to the machine, the machine will stop after a finite number of steps, see [19, Def. 3.3]. It is termed *undecidable* otherwise. A Turing machine is a deterministic automaton with an infinite number of states, see [19, Def. 3.1]. There are questions for Turing machines which are undecidable.

Decidability questions in control and system theory have been dealt with by several authors, see [3, 20, 21, 23, 27]. But the development of control and system theory requires a broader and deeper application of the concepts and results of decidability and of complexity. In control of discrete-event systems the concept of decidability deserves a place as prominent as finite-dimensionality in the system theory of linear systems. If a problem is decidable then computational properties of the problem may be used to delimit further the class of discrete-event systems for which theory can be developed. Approaches like hierarchical modeling and modularity are also used to combat complexity. But it is far from clear how to apply hierarchical modeling in specific situations.

In the research area of hybrid systems much attention has been given both by computer scientists and by control theorists to equivalences. R. Alur and D. Dill, see [1], identified a class of timed automata of which the untimed language is equivalent to a language of a finite-state automaton. This result may be considered as a solution to an extension of the problem formulated above. In the extension the first system is a timed-automaton and the map p_1 projects away the time component of the timed-event sequence. This approach has motivated much research in hybrid systems, see for example [12]. It seems of interest to also explore other equivalence relations and other classes of equivalent systems.

3 Extensions of the problem

The problem may be extended in several directions of which a few are described below.
The problem above was formulated for a deterministic automaton. Other classes of discrete-event systems for which the problem is of interest include nondeterministic automata, see [19], Petri nets, see [9], and process algebras, see [2]. Chomsky, [6], has introduced a hierarchy of languages according to which the classes of languages generated by finite-state automata, by Petri nets, and by process algebras are strictly contained in each other respectively. There exist subclasses of Petri nets for which certain control problems are decidable and there exists other such classes for which these are undecidable. The case of a decidable problem is an example of a solution to the proposed problem formulated above. Which conditions on the subclass and on the problem result in decidability of the problem? Yet another extension is to consider infinite string languages represented by finite state automata, see [24, 25, 26].
A possibly fruitfull extension is to consider for nondeterministic automata or for transition systems other concepts of equivalence as treated in the area of semantics of computer science. For nondeterministic systems failure semantics, see [15], failure trace semantics, see [18], and bisimulation equivalence, see [14], are used. Bisimulation is the finest of these relations, see [2]. Sources on bisimulation are [2, 11, 14]. For control problems more explicit control induced semantics may be required. The semantics may be structured by the underlying control system and by the problem, such as the hierarchical or decentralized structure of the system.
For the class of discrete-event systems A_1 for which the required equivalence does not exist, a further decomposition may be of interest to theory development. Do their exist subclasses of that class of systems which are equivalent with other classes of discrete-event systems having specified computational and complexity properties? In computer science a Turing machine is used to define decidability. This approach to computation and complexity started with the work of A. Church, see [7]. Would it be useful for complexity theory to relate complexity concepts to another automaton or to another discrete-event system? If parallel processing is allowed, then other properties of the second system may be considered.
Finally, the extension of the problem to equivalences of hybrid systems is of interest. Research is required into useful equivalence relations and their equivalence classes. For a paper on this approach see [8]. The concept of

abstraction is used for an equivalence relation in which part of the trajectory set is projected away, see [1, 12]. A paper by the author of this chapter for reachability of hybrid control systems, see [28], is based on the same approach. For hybrid systems the relation between discrete objects and real vector spaces needs to be exploited further. For an entry into this point the following references are recommended [4, 5].
Decidability of stability and of controllability of hybrid systems is discussed in the paper [3] but stability is not part of the problem proposed in this chapter.

4 References

[1] R. Alur and D.L. Dill. A theory of timed automata. *Theoretical Computer Science*, 126:183 - 235, 1994.

[2] J.C.M. Baeten and W.P. Weijland. *Process algebra.* Cambridge University Press, Cambridge, 1990.

[3] V. Blondel and J. Tsitsiklis. Complexity of stability and controllability of elementary hybrid systems. To appear in *Automatica.* Report LIDS-P 2388, LIDS, Massachusetts Institute of Technology, Cambridge, MA, 1997.

[4] L. Blum, F. Cucker, M. Shub, and S. Smale. *Complexity and real computation.* Springer, New York, 1997.

[5] L. Blum, M. Shub, and S. Smale. On a theory of computation and complexity over the real numbers. *Bull. Amer. Math. Soc.*, 21:1–46, 1989.

[6] N. Chomsky. Three models for the description of language. *IRE Trans. Information Theory*, 2:113–124, 1956.

[7] A. Church. Logic, arithmetic, and automata. In *Proc. 1962 Internat. Congress of Math.*, pages 15–22, X, 1963. Mittag-Leffler Inst.

[8] B. Dasgupta and E.D. Sontag. A polynomial-time algorithm for an equivalence problem which arises in hybrid system theory. In *Proceedings of the Conference on Decision and Control*, page to appear, New York, 1998. IEEE Press.

[9] R. David and H. Alla. *Petri nets and grafcet.* Prentice Hall, New York, 1992.

[10] S. Eilenberg. *Automata, languages, and machines (Volumes A and B).* Academic Press, New York, 1974, 1976.

[11] T.A. Henzinger. Hybrid automata with finite bisimulations. In Z. Fülöp and F. Géseg, editors, *ICALP95: Automata, languages, and programming*, pages 324–335, Berlin, 1995. Springer-Verlag.

[12] T.A. Henzinger, P.W. Kopke, A. Puri, and P. Varaiya. What's decidable about hybrid automata. In X, editor, *Proceedings of the 27th Annual Symposium on Theory of Computing*, pages 373–382, X, 1995. ACM Press.

[13] K. Krohn and J. Rhodes. Algebraic theory of machines. I. Prime decomposition theorem for finite semigroups and machines. *Trans. Amer. Math. Soc.*, 116:450–464, 1965.

[14] R. Milner. *Communication and concurrency.* Prentice-Hall, Englewood Cliffs, NJ, 1989.

[15] A. Overkamp. Supervisory control using failure semantics and partial specifications. *IEEE Trans. Automatic Control*, 42:498–510, 1997.

[16] P.J.G. Ramadge and W.M. Wonham. The control of discrete event systems. *Proc. IEEE*, 77:81–98, 1989.

[17] G. Rozenberg and A. Salomaa, editors. *Handbook of formal languages, Volumes 1-3.* Springer, Berlin, 1997.

[18] M. Shayman and R. Kumar. Supervisory control of nondeterministic systems with driven events via prioritized synchronization and trajectory models. *SIAM J. Control & Opt.*, 33:469–497, 1995.

[19] M. Sipser. *Introduction to the theory of computation.* PWS Publishing Company, Boston, 1997.

[20] E.D. Sontag. On certain questions of rationality and decidability. *J. Comp. Syst. Sci.*, 11:375–381, 1975.

[21] E.D. Sontag. Controllability is harder to decide than accessibility. *SIAM J. Control & Opt.*, 26:1106–1118, 1988.

[22] E.D. Sontag. *Mathematical control theory: Deterministic finite dimensional systems.* Springer-Verlag, New York, 1990.

[23] E.D. Sontag. From linear to nonlinear: Some complexity comparisons. In *Proc. IEEE Conference on Decision and Control*, pages 2916–2920, New York, 1995. IEEE Press.

[24] L. Staiger. ω-languages. In G. Rozenberg and A. Salomaa, editors, *Handbook of formal languages*, pages 339–387. Springer, Berlin, 1997.

[25] J.G. Thistle and W.M. Wonham. Control of infinite behavior of finite automata. *SIAM J. Control & Opt.*, 32:1075–1097, 1994.

[26] W. Thomas. Automata on infinite objects. In J. van Leeuwen, editor, *Handbook of theoretical computer science, Volume B*, pages 133 - 191. Elsevier Science Publishers B.V., Amsterdam, 1990.

[27] J.N. Tsitsiklis. On the control of discrete-event systems. *Math. Control Signals Systems*, 2:95–107, 1989.

[28] J.H. van Schuppen. A sufficient condition for controllability of a class of hybrid systems. In T.A. Henzinger and S. Sastry, editors, *Hybrid systems: Computation and control*, number 1386 in Lecture Notes in Computer Science, pages 374–383, Berlin, 1998. Springer.

[29] K.C. Wong and W.M. Wonham. Hierarchical control of discrete-event systems. *J. Discrete Event Dynamics Systems*, 6:241–273, 1996.

[30] Sheng Yu. Regular languages. In G. Rozenberg and A. Salomaa, editors, *Handbook of formal languages, Volume 1*, pages 41–110. Springer, Berlin, 1997.

[31] H. Zhong and W.M. Wonham. On the consistency of hierarchical supervisors in discrete-event systems. *IEEE Trans. Automatic Control*, 35:1125–1134, 1990.

48

Covering numbers for input-output maps realizable by neural networks

M. Vidyasagar

Centre for Artificial Intelligence and Robotics
Raj Bhavan Circle, High Grounds
Bangalore 560 001
INDIA
sagar@cair.res.in

1 Problem Statements

We begin by introducing some standard terminology, namely covering numbers, and neural architectures.

First the notion of covering numbers is introduced. Suppose Y is a set and that ρ is a pseudometric on Y. Recall that the only difference between a metric and a pseudometric is that if ρ is a pseudometric, then $\rho(x, y) = 0$ *need not* necessarily imply that $x = y$. However, all the standard notions of convergence, open and closed sets, continuity, etc. hold in pseudometric spaces. Now suppose $S \subseteq Y$, and $\epsilon > 0$ is a given number. A collection of elements $s_1, \ldots, s_n \in S$ is said to be an **ϵ-cover** of S if, for every $s \in S$, there exists an index i such that $\rho(s, s_i) \leq \epsilon$. In other words, the union of the closed balls of radius ϵ centered at each of the s_i contains S. The **ϵ-covering number** of S is defined as the *smallest* number n such that there exists an ϵ-cover for S of cardinality n, and is denoted by $N(\epsilon, S, \rho)$.

A very natural pseudometric of interest to us is the following: Suppose X is a set, $\mathcal{H}$ is a family of (measurable) functions mapping X into $[0, 1]$, and P is a probability measure on X. Given two functions $f, g \in \mathcal{H}$, define

$$d_P(f, g) := \int_X |f(x) - g(x)| \, P(dx).$$

Then $d_P(\cdot, \cdot)$ is a pseudometric on $\mathcal{H}$.

It is a standard result in topology that the covering number $N(\epsilon, S, \rho)$ is finite for every $\epsilon > 0$ if and only if the closure of S is a compact set. Such a set is called **totally bounded**.

Next, we define precisely what is meant by a neural architecture. A **neural network** is a labelled acyclic directed graph with the following features:

- The graph has at least one node with fan-in zero. The set of nodes with fan-in zero is called the set of **input nodes** of the network. Similarly, the graph has at least one node with fan-out zero. The set of nodes with zero fan-out is called the set of **output nodes** of the network.

- Every node of the graph, except for the input nodes, is labelled with a suitable input-output mapping. Thus, if node i has fan-in r_i, then the label of node i is a mapping $g_i : \Re^{r_i} \rightarrow \Re$, chosen from some prescribed *family* of mappings $\mathcal{G}_i$.

The functioning of the neural network is as follows: Let k denote the number of input nodes, and let $X \subseteq \Re^k$ be a given set, called the **input set.** Whenever a vector $\mathbf{x} := [x_1 \dots x_k]^t \in X$ is assigned to the input nodes, every subsequent node i applies the mapping g_i to its inputs. Since the graph is acyclic, this procedure is well-defined. By proceeding in this sequential fashion, one can eventually assign values to each node of the network, and in particular, to each of the output nodes. In this way, the neural network defines a map $h : X \rightarrow \Re^n$, where n is the number of output nodes. Thus, in the present formulation, a neural network consists of: (i) an acyclic directed graph, (ii) an input set $X \subseteq \Re^k$, and (iii) a label g_i for each node i other than the input nodes, where g_i is chosen from a prespecified family $\mathcal{G}_i$. The **architecture** of the neural network consists of its graph, its input set X, and the *families* of maps $\mathcal{G}_i$. Each choice of g_i from $\mathcal{G}_i$ leads to a mapping h from X to $\Re^n$. By *varying* each g_i over the corresponding set $\mathcal{G}_i$, one generates a *family* $\mathcal{H}$ of mappings from X into $\Re^n$.
The open problems stated here fall into two categories:

1. When are certain families of functions $\mathcal{H}$ totally bounded with respect to the pseudometric d_P described above?

2. If $\mathcal{H}$ is totally bounded, is it possible to derive *tight* bounds for the covering number $N(\epsilon, \mathcal{H}, d_P)$ for each ϵ, where P is a specified probability measure on X?

Now for the problems themselves:
Problem 1: Suppose X is a compact subset of $\Re^k$ for some integer k; for instance, take $X = [-1, 1]^k$. By suitable scaling, many problems can be put into this framework. If each neuron implements a set of piecewise polynomial input-output relationships, then it is known (see below) that the corresponding family $\mathcal{H}$ is totally bounded, for *every* probability measure P. But the available bounds on the covering numbers $N(\epsilon, \mathcal{H}, d_P)$ are far too conservative. The question is: Can one improve on these bounds in the case where P is the uniform probability measure, or some other "reasonable" probability measure (e.g., a nonatomic measure whose density function is separable and continuous)?

Problem 2: There are some nice examples of function families $\mathcal{H}$ that are *not* totally bounded for *every* probability measure P, but are nevertheless totally bounded if P is the *uniform* probability measure. For example, suppose X is a *convex* subset of $\Re^k$, and that $\mathcal{H}$ is a collection of *concave* functions. Under these conditions, the sous-graph

$$s(h) := \{(x, y) : 0 \leq y \leq h(x)\}$$

is a convex set for each $h \in \mathcal{H}$. Then $\mathcal{H}$ is totally bounded if P is the uniform measure, but not in general. Now, if the sous-graph of every function in $\mathcal{H}$ is a polyhedral set, then such a family can be realized using only perceptrons and a single "and" gate. It can be shown that $\mathcal{H}$ is *not* totally bounded for *every* probability measure P. The reason for the noncompactness of this family is that the number of perceptrons is *not* bounded a priori, and could depend on the function h. On the other hand, it can be shown through some highly technical arguments that if P is the uniform probability, or if P has a separable density function, i.e., the density function has the form

$$p(x_1, \ldots, x_k) = \prod_{i=1}^{k} p_i(x_i)$$

where each p_i is continuous, then $\mathcal{H}$ is indeed totally bounded with respect to d_P. Is it possible to discover other "natural" classes of function families $\mathcal{H}$ that have this "conditional total boundedness" property?

2 Motivation

There are two rather disparate-looking reasons why one would want to study this type of problem.

Model-Free Learning Using Neural Networks: The main motivation for using neural networks is their alleged ability to "generalize." At an abstract level, one can state the problem as follows: Suppose one is given a neural network architecture, which as we have seen above amounts to specifying a family $\mathcal{H}$ of input-output mappings. Suppose one is given a set of input-output data $(x_1, y_1), \ldots, (x_m, y_m)$. The idea is to "train" a network of the given architecture on these input-output pairs in such a way that, if the network is then shown a randomly selected "testing" input-output pair (x, y), its output value $h(x)$ is as close as possible to the "correct" output y. This is equivalent to the problem of fitting randomly generated data by an input-output mapping belonging to the family $\mathcal{H}$. A very natural "training" algorithm is to choose a network, among all networks of the specified architecture, that fits the *training data* as well as possible. In other words, the trained network minimizes the fitting error to the data among all the networks of the specified architecture. This is the most commonly used training philosophy in neural networks. Methods such as back-propagation are merely attempts at *implementing* this particular philosophy. The objective of the training is to ensure that, as the number of data points approaches infinity, the "trained" network in fact approaches an "optimal" network, i.e., a network with the property that the expected value of $|y - h(x)|$ is as small as possible as the function h varies over $\mathcal{H}$. The adjective "model-free" in the name of the problem means that it is *not* assumed that the minimum value is zero. Equivalently, there need not exist a "true" function $g \in \mathcal{H}$ such that $y_i = g(x_i)\ \forall i$.

It is known that the trained network converges to an optimal network provided the family $\mathcal{H}$ of input-output mappings realized by the architecture is totally bounded with respect to the pseudometric d_P, where P is the probability measure that generates the data (see [12], Section 6.6). Thus it becomes of interest to ascertain when this total boundedness holds. More generally, suppose one has an option of trying out *several different* neural network architectures, and is interested in choosing an "optimal" neural network architecture to fit a given set of data. By now there is a great deal of experimental evidence to show that, if the network architecture is too "rich," then the network will perform extremely well on the training data, but do very poorly on previously unseen testing data. In effect, the network merely "memorizes" the data, and has no generalization ability; this problem is often referred to as "over-fitting." On the other hand, if the network architecture is not rich enough, then the performance of the network will be poor even on the training data. This trade-off is similar to the bias-variance trade-off in identification. There is a method known as "structural risk minimization" that *quantifies* this trade-off (see [11]). In order for this method to be meaningful, it is essential to have reasonably good estimates for the covering numbers $N(\epsilon, \mathcal{H}, d_P)$. If one has a *nested collection* of families $\{\mathcal{H}_i\}$ where $\mathcal{H}_i \subseteq \mathcal{H}_{i+1}$, and if *all* estimates $N(\epsilon, \mathcal{H}_i, d_P)$ are of roughly the same degree of conservativeness, then the method of structural risk minimization seems to work reasonably well.

Randomized Algorithms for Robust Controller Synthesis: By now it is known that several problems in robust controller synthesis are NP-hard (see e.g. [2]). Thus one is forced to change one's notion of a "solution" to a problem in such cases. An approach that is recently gaining popularity is that of *randomized* algorithms. It turns out that, in many robust control problems, the underlying family $\mathcal{H}$ *is* totally bounded, for *every* probability measure P. However, the estimates for $N(\epsilon, \mathcal{H}, d_P)$ are far too conservative, which means that the prior estimates for the number of random plant-controller pairs that one needs to generate is impractically large. Numerical tests indicate that these estimates are too conservative by two or three orders of magnitude. It would be desirable to get better estimates so as to make the randomized approach practically feasible.

3 Available Results

The above problems as well as the requisite background material are discussed in great detail in the recently published book [12]. Only some relevant results are quoted here.

The central concept is that of the Pollard, or pseudo-dimension of a function family. The precise definition is given next.

Definition ([12], p. 74): Let X be a given set, and let $\mathcal{H}$ be a family of measurable functions mapping X into $[0,1]$. A set $S = \{x_1, \ldots, x_n\} \subseteq X$ is said to be **P-shattered** by $\mathcal{H}$ if there exists a real vector $\mathbf{c} \in [0,1]^n$ such that, for every binary vector $\mathbf{e} \in \{0,1\}^n$, there exists a corresponding function $f_\mathbf{e} \in \mathcal{H}$ such that

$$f_\mathbf{e}(x_i) \geq c_i \text{ if } e_i = 1, \text{ and } f_\mathbf{e}(x_i) < c_i \text{ if } e_i = 0.$$

The **P-dimension** of $\mathcal{H}$, denoted by P-dim$(\mathcal{H})$, is defined as the largest integer n such that there exists a set of cardinality n that is P-shattered by $\mathcal{H}$.

The significance of the P-dimension is brought out in the following result:

Theorem ([5], [12], Corollary 4.2, p. 83): Suppose the family $\mathcal{H}$ has finite P-dimension, and suppose P-dim$(\mathcal{H}) \leq d$. Then $\mathcal{H}$ is totally bounded with respect to d_P for *every* probability measure P. Moreover,

$$N(\epsilon, \mathcal{H}, d_P) \leq 2 \left(\frac{2e}{\epsilon} \ln \frac{2e}{\epsilon} \right)^d.$$

In the case of mappings that map X into the binary set $\{0,1\}$, the P-dimension is replaced by the VC- (Vapnik-Chervonenkis) dimension. Though the above result is quite an intellectual *tour de force*, the estimate contained in it is rather conservative, especially because it does not in any way depend on the *particular* probability measure P.

The above theorem shows that it is desirable to have estimates for the P-dimension of specific families of functions $\mathcal{H}$. One such estimate, which is applicable to broad classes of neural networks and also several problems in robust control, is the following:

Theorem (cf. [6, 7, 12], Corollary 10.2, p. 330): Suppose the collection of functions $\mathcal{H}$ mapping X into $[0,1]$ has the following properties:

- X is a subset of some Euclidean space $\Re^k$.
- Every function $h \in \mathcal{H}$ has the form $h(x) = g(x,y)$ for some $y \in Y \subseteq \Re^l$, where g is a *fixed* function mapping $X \times Y \to [0,1]$. In other words, the collection $\mathcal{H}$ is generated by *varying* y over the set $Y \subseteq \Re^l$.

- Suppose that, for every $c \in [0,1]$, the inequality $g(x,y) - c > 0$ can be written as a *Boolean formula* involving s "atomic" inequalities $\tau_1(x,y,c) > 0, \ldots, \tau_s(x,y,c) > 0$, where each $\tau_i(x,y,c)$ is a polynomial in x, y, c, and moreover, the degree of τ_i with respect to y, c is no larger than r for each i.

Under these conditions,

$$\text{P-dim}(\mathcal{H}) \leq 2l \lg(4esr).$$

The above result is quite deep, and involves the use of some sophisticated methods from algebraic topology. Moreover, it is not too conservative an estimate for the P-dimension. Thus the main source of conservatism is in estimating the covering numbers in terms of the P-dimension.
It can be noticed that the various bounds given above are extremely general. In particular, (i) they are "distribution-free" in the sense that the underlying probability measure P does not affect the bounds, and (ii) the nature of the neural network also does enter into the bounds, except for the total number of elements. If one wishes to improve the above bounds, it is necessary to take advantage of the structure of the underlying problem. In the case of mappings that are realizable by recurrent perceptron networks, it is possible to obtain improved bounds for the VC-dimension; see for example [3, 4, 8, 10]. Other approaches that are *not* based on VC-dimension type arguments can be found in [9, 1].

4 References

[1] Peter L. Bartlett, "The sample complexity of pattern classification with neural networks: The size of the weights is more important than the size of the network," *IEEE Trans. Inf. Thy*, Vol. 44, No. 2, pp. 525-536, March 1998.

[2] V. Blondel and J. N. Tsitsiklis, "NP-hardness of some linear control design problems," *SIAM J. Control and Opt.*, 35, pp. 2118-2127, 1997.

[3] B. Dasgupta and E.D. Sontag, "Sample complexity for learning recurrent perceptron mappings," summary in *Advances in Neural Information Processing*, 8, MIT Press, Cambridge, MA, pp. 204-210, 1996.

[4] B. Dasgupta and E.D. Sontag, "Sample complexity for learning recurrent perceptron mappings," *IEEE Trans. Info. Thy*, Vol. 42, pp. 1479-1487, 1996.

[5] D. Haussler, "Decision theoretic generalizations of the PAC model for neural net and other learning applications," *Information and Computation*, 100, pp. 78-150, 1992.

[6] M. Karpinski and A.J. Macintyre, "Polynomial bounds for VC dimension of sigmoidal neural networks," *Proc. 27th ACM Symp. Thy. of Computing*, pp. 200-208, 1995.

[7] M. Karpinski and A.J. Macintyre, "Polynomial bounds for VC dimension of sigmoidal and general Pfaffian neural networks," *J. Comp. Sys. Sci.*, (to appear).

[8] P. Koiran, and E.D. Sontag, "Vapnik-Chervonenkis dimension of recurrent neural networks," *Discrete Applied Math.*, 1998, to appear. (Summary in *Proceedings of Third European Conference on Computational Learning Theory*, Jerusalem, March 1997, Springer Lec. Notes in Computer Science 1208, pages 223-237.)

[9] A. Kowalczyk, H. Ferra and J. Szymanski, "Combining statistical physics with VC-bounds on generalisation in learning systems," *Proceeding of the Sixth Australian Conference on Neural Networks (ACNN'95*,pp. 41-44, Sydney, 1995.

[10] Sontag, E.D., "A learning result for continuous-time recurrent neural networks," Systems and Control Letters, 1998 (to appear).

[11] V.N. Vapnik, *Estimation of Dependences Based on Empirical Data*, Springer-Verlag, 1982.

[12] M. Vidyasagar, *A Theory of Learning and Generalization: With Applications to Neural Networks and Control Systems*, Springer-Verlag, London, 1997.

49

A powerful generalization of the Carleson measure theorem?

George Weiss

Centre for Systems and Control Engineering
School of Engineering
University of Exeter
Exeter EX4 4QF
UNITED KINGDOM
G.Weiss@exeter.ac.uk

1 Description of the problem

First we describe the structure of an infinite-dimensional linear system with a scalar observation (measurement).
We consider a system whose state space is a complex Hilbert space X. The evolution of the state is governed by a strongly continuous semigroup of operators $\mathbf{T}$, so that $\mathbf{T}_t \in \mathcal{L}(X)$ for all $t \geq 0$. If the initial state of the system is $x_o \in X$, then its state trajectory x is given by $x(t) = \mathbf{T}_t\, x_o$. We denote by A the infinitesimal generator of $\mathbf{T}$ and by $D(A)$ its domain. Assume that a linear functional $C : D(A) \to \mathbb{C}$ is given such that

$$|Cz| \leq k\left(\|z\| + \|Az\|\right), \tag{49.1}$$

for some $k > 0$ (i.e., C is A-bounded). C may be regarded as a densely defined functional on X. If C has a bounded extension to X, then we call it *bounded*. Otherwise, it is called *unbounded* on X and in this case it must be non-closable (this is easy to check).
We assume that for each $x_o \in D(A)$ this system generates a scalar output signal y defined by $y(t) = Cx(t)$. For this reason, C is called the *observation operator* of the system. It is clear that y is continuous. We are interested to know if the following estimate holds:

$$\|y\|_{L^2[0,\infty)} \leq m_o\|x_o\|, \tag{49.2}$$

for some $m_o \geq 0$. If (49.2) holds, we call C *infinite-time admissible*. It is clear that this property holds if and only if the operator from $D(A)$ to

$C[0,\infty)$ defined by $y(t) = C\mathbf{T}_t x_o$ has a bounded extension to an operator $\Psi : X \to L^2[0,\infty)$. The existence of such an extension is crucial in many system-theoretic considerations, including Lyapunov equations, linear quadratic optimal control and H^∞ control, see for example [5], [11], [12], [15].
Infinite-time admissibility of an observation operator is a slightly stronger concept than *admissibility*, which is defined by replacing in (49.2) $L^2[0,\infty)$ by $L^2[0,\tau]$, for some $\tau > 0$. The choice of τ does not matter, all values are equivalent. A bounded C is always admissible, but it may not satisfy (49.2). If the semigroup is exponentially stable, then infinite-time admissibility is equivalent to admissibility. However, in this note we make no stability assumptions on $\mathbf{T}$. Admissibility of observation operators has been defined and studied in the operator-theoretic context in [6], [10], [13] and several other papers, see the references in the papers cited. Infinite-time admissibility has been studied in [2], [3], [5] and the references therein. We refer the reader to these papers for more background and results on the subject.
If we apply the Laplace transformation to y, we obtain that for $\operatorname{Re} s$ sufficiently large $\hat{y}(s) = C(sI - A)^{-1}x_o$. If C is infinite-time admissible, then $\hat{y} \in H^2(\mathbb{C}_+)$ ($\mathbb{C}_+$ denotes the open right half-plane), so that in particular $|\hat{y}(s)| \leq \|y\|_{L^2[0,\infty)}/\sqrt{2\operatorname{Re} s}$. Combining these facts with (49.2), we get the estimate

$$\|C(sI - A)^{-1}\|_{\mathcal{L}(X,\mathbb{C})} \leq \frac{m}{\sqrt{\operatorname{Re} s}}, \quad \forall\, s \in \mathbb{C}_+ , \tag{49.3}$$

where $m = m_o/\sqrt{2}$. Well, actually the left-hand side of (49.3) is not necessarily defined on all of $\mathbb{C}_+$, because A might have some spectrum there. However, if (49.2) holds then $C(sI - A)^{-1}$ must have an analytic continuation to $\mathbb{C}_+$ and the estimate must hold with this extension. We conjecture that the converse implication is also true:

Conjecture. (49.3) *implies that C is infinite-time admissible.*

When we say that A and C satisfy (49.3), we mean that an analytic continuation of $C(sI - A)^{-1}$ exists for which (49.3) holds. It is interesting to note that the conjecture does not become at all simpler if we restrict ourselves to bounded A and C (see Remark 3.7 in [10]).

2 How is the conjecture related to the Carleson measure theorem?

We introduce a notation for certain rectangles in the closed right half-plane $\mathbb{C}_{cl}$ (the closure of $\mathbb{C}_+$). For any $h > 0$ and any $\omega \in \mathbb{R}$,

$$R(h,\omega) = \{z \in \mathbb{C} \mid 0 \leq \operatorname{Re} z < h,\ \omega - h \leq \operatorname{Im} z < \omega + h\}.$$

A positive measure μ on (the Borel subsets of) $\mathbb{C}_{cl}$ is called a *Carleson measure* if there is an $M \geq 0$ such that for all h and ω as above,

$$\mu(R(h,\omega)) \leq M \cdot h. \tag{49.4}$$

The *Carleson measure theorem* says that if μ satisfies (49.4), then denoting $c = 20\sqrt{\frac{M}{2\pi}}$, we have that for all $f \in H^2(\mathbb{C}_+)$,

$$\int_{\mathbb{C}_{cl}} |f|^2 \, d\mu \leq c^2 \cdot \|f\|^2_{H^2}. \tag{49.5}$$

The above definition of Carleson measures and the statement of the theorem are a bit unorthodox, because here μ is a measure on $\mathbb{C}_{cl}$, whereas in the standard references [6], [7] μ is a measure on the open half-plane $\mathbb{C}_+$. (The above formula for c follows from Section 3 of [4]. More often, the theorem is stated in its disk version, where μ is a finite measure on the open unit disk, see [1], [8].) This slight generalization in the domain of μ is needed for our purposes, but it requires a small explanation: it is easy to see from (49.4) that the restriction of μ to the imaginary axis $i\mathbb{R}$ is absolutely continuous with respect to the (one-dimensional) Lebesgue measure, and its Radon-Nikodym derivative is in $L^\infty(i\mathbb{R})$. Hence, the integral of the boundary trace of $|f|^2$ (which is in $L^1(i\mathbb{R})$) with respect to this part of μ is well defined. The extension of the Carleson measure theorem from the open half-plane to the closed one is very easy.

Taking in (49.5) $f(z) = 1/(s+z)$ with $\operatorname{Re} s > 0$, we obtain that with $m^2 = c^2\pi$,

$$\int_{\mathbb{C}_{cl}} \frac{d\mu(z)}{|s+z|^2} \leq \frac{m^2}{\operatorname{Re} s}, \qquad \forall\, s \in \mathbb{C}_+. \tag{49.6}$$

It is not difficult to show that (49.6) implies (49.4) with $M = 5m^2$. Thus, (49.4), (49.5) and (49.6) are equivalent (a possible value for any one of M, c and m determines possible values for the other two), see [2], [8], [14].

Now let $\mathbf{T}$, A and C be as in Section 1 and assume that $\mathbf{T}_t$ is normal. We claim that the conjecture from Section 1 is true for such semigroups:

Theorem 19 *If* $\mathbf{T}_t$ *is normal, then* (49.3) *implies* (49.2).

Proof. Following the notation and terminology from [9], we have

$$\mathbf{T}_t = \int_{\mathbb{C}} e^{-zt} dE(z), \qquad (\bar{s}I - A^*)^{-1} = \int_{\mathbb{C}} \frac{1}{\bar{s}+\bar{z}} \, dE(z), \tag{49.7}$$

where E is the spectral decomposition of $-A$. From the last formula, taking $b \in X$ and denoting by $E_{b,b}$ the positive measure on $\mathbb{C}$ defined by $E_{b,b}(S) = \langle E(S)b, b\rangle = \|E(S)b\|^2$, we get that

$$\|(\bar{s}I - A^*)^{-1}b\|^2 = \int_{\mathbb{C}} \frac{1}{|s+z|^2} \, dE_{b,b}(z).$$

First we assume that C is bounded, i.e., $C \in X'$. Taking in the above formula $b = C^*$ and denoting $\mu(S) = \|CE(S)\|^2 = E_{b,b}(S)$, we obtain

$$\|C(sI - A)^{-1}\|^2 = \int_{\mathbb{C}} \frac{d\mu(z)}{|s+z|^2} \, . \tag{49.8}$$

If (49.3) holds, then (49.8) implies that $\mu(S) = 0$ if $S \cap \mathbb{C}_{cl} = \emptyset$, and also that on $\mathbb{C}_{cl}$, μ satisfies (49.6), so that it is a Carleson measure. Take $x_o \in D(A)$ and put $y(t) = C\mathbf{T}_t x_o$. Then $\|y\|_{L^2[0,\infty)} = \sup_{\|v\|_{L^2} \leq 1} \left| \int_0^\infty C\mathbf{T}_t x_o \overline{v}(t) dt \right|$. Using (49.7), we see that with $b = C^*$

$$\begin{aligned} \int_0^\infty C\mathbf{T}_t x_o \overline{v}(t) dt &= \int_0^\infty < x_o, \int_{\mathbb{C}_{cl}} e^{-\overline{z}t} v(t) dE(z) b > dt \\ &= < x_o, \int_{\mathbb{C}_{cl}} \hat{v}(\overline{z}) dE(z) b > , \end{aligned}$$

whence

$$\|y\|^2_{L^2[0,\infty)} \leq \sup_{\|v\|_{L^2} \leq 1} \|x_o\|^2 \cdot \int_{\mathbb{C}_{cl}} |\hat{v}(\overline{z})|^2 d\mu(z) \, .$$

Now from (49.5) with $f(z) = \overline{\hat{v}(\overline{z})}$, we get that $\|y\|_{L^2[0,\infty)} \leq c\sqrt{2\pi} \cdot \|x_o\|$, i.e., (49.2) holds with $m_o = c\sqrt{2\pi}$.
Now we drop the boundedness assumption on C, but (49.1) and (49.3) still hold. It is easy to see that for any bounded Borel set $S \subset \mathbb{C}$, $E(S)$ is a bounded operator from X to $D(A)$ with the graph norm (the right-hand side of (49.1)). We introduce a sequence of bounded approximations to C by $C_n = CE(R_n)$, where R_n is the square obtained as the union of $R(n,0)$ and $-R(n,0)$. Since (49.3) holds for C_n in place of C, with the same m, by the earlier argument for bounded C we obtain that

$$\int_0^\infty |C_n \mathbf{T}_t x_o|^2 \, dt \leq m_o^2 \, \|x_o\|^2 \, , \qquad \forall x_o \in D(A) \, ,$$

with m_o independent of n. Since $C_n \mathbf{T}_t x_o \to C\mathbf{T}_t x_o$ uniformly on $[0, \infty)$, the last estimate implies (49.2). ■

With the above notation, allowing unbounded C, we can define $\mu(S) = \|CE(S)\|^2$ (as for bounded C), and μ extends to a positive measure on $\mathbb{C}$, which is a Carleson measure iff C is infinite-time admissible.
Note that we may have $\mu(i\mathbb{R}) > 0$. For example, if $\mathbf{T}$ is the left shift semigroup on $L^2(\mathbb{R})$ and $Cx = x(0)$, then C is infinite-time admissible and μ is simply $\frac{1}{2\pi}$ times the Lebesgue measure on $i\mathbb{R}$.
It is not difficult to derive the Carleson measure theorem from Theorem 19. This shows that Theorem 19 is (in a certain meta-mathematical sense) equivalent to the Carleson measure theorem. (In the strictly logical sense this is trivial, any two true statements are equivalent.) This shows that the conjecture, if it were true, would constitute a powerful generalization of the

Carleson measure theorem, since arbitrary operator semigroups are more general than normal semigroups.
What reason do we have to hope that such a generalization might be true? The most important reason, as I see it, is that the conjecture is true for exponentially stable and right invertible semigroups. This follows from a more general result in Section 4 of [14]. For intersting related results we refer to [2], [3], [4], [5], [10].

3 References

[1] P.L. Duren, *Theory of H^p Spaces*, Academic Press, New York, 1970.

[2] P. Grabowski, "Admissibility of observation functionals", *Internat. J. of Control* **62**, pp. 1161–1173 (1995).

[3] P. Grabowski and F. Callier, "Admissible observation operators. Semigroup criteria of admissibility", *Integral Equations and Operator Theory* **25**, pp. 182–198 (1996).

[4] S. Hansen and G. Weiss, "The operator Carleson measure criterion for admissibility of control operators for diagonal semigroups on l^2", *Systems and Control Letters* **16**, pp. 219–227 (1991).

[5] S. Hansen and G. Weiss, "New results on the operator Carleson measure criterion", *IMA J. of Math. Control & Inform.* **14**, pp. 3–32 (1997).

[6] L.F. Ho and D.L. Russell, "Admissible input elements for systems in Hilbert space and a Carleson measure criterion", *SIAM J. Control & Optim.* **21**, pp. 614–640 (1983).

[7] P. Koosis, *Introduction to H^p Spaces*, Cambridge University Press, Cambridge, 1980.

[8] N.K. Nikolskii, *Treatise on the Shift Operator*, Grundlehren der math. Wissenschaften, Springer-Verlag, Berlin, 1986.

[9] W. Rudin, *Functional Analysis*, McGraw-Hill, New York, 1973.

[10] D.L. Russell and G. Weiss, "A general necessary condition for exact observability", *SIAM J. Control & Optim.* **32**, pp. 1–23 (1994).

[11] O. Staffans, "Quadratic optimal control of stable well-posed linear systems", *Trans. of the Amer. Math. Soc.*, to appear.

[12] O. Staffans, "Feedback representations of critical controls for well-posed linear systems", submitted in 1997.

[13] G. Weiss, "Admissibility observation operators for linear semigroups", *Israel J. Math.* **65**, pp. 17–43 (1989).

[14] G. Weiss, "Two conjectures on the admissibility of control operators", in *Estimation and Control of Distributed Parameter Systems*, eds: W. Desch, F. Kappel, K. Kunisch, Birkhäuser-Verlag, Basel, 1991.

[15] M. Weiss and G. Weiss, "Optimal control of stable weakly regular linear systems", *Math. Control, Signals and Systems*, to appear.

50
Lyapunov theory for high order differential systems

Jan C. Willems

Mathematics Institute
University of Groningen
P.O. Box 800
9700 AV Groningen
THE NETHERLANDS
willems@math.rug.nl

1 Introduction

One of the noticeable aspects of the theory of dynamics is the prevalence, since the work of Lyapunov and Poincaré, of first order differential equations. Of course, there are good conceptual and theoretical reasons for this, connected with the notion of state and the specification of initial conditions. Since the late fifties also areas as control theory have relented to this point of view. This introduction of state models in fact brought with it a renaissance in this field.

Nevertheless, models obtained from first principles will seldomly be in the form of a system of first order differential equations in which the time rate of change of the state is given as an explicit function of the instantaneous value of the state and the input. As such, it is important to develop theory and algorithms that pass directly from first principle models to results and algorithms as, for example, stability conclusions or controller specifications.

In this note we pose the problem of stability and of the construction of Lyapunov functions in the setting of high order differential equations.

2 Linear Theory

Let $R \in \mathbb{R}^{\bullet \times q}[\xi]$ be a polynomial matrix and consider the dynamical system

$$\Sigma_R := (\mathbb{R}, \mathbb{R}^q, \mathcal{B}_R) \tag{50.1}$$

with

$$\mathcal{B}_R := \{w \in C^\infty(\mathbb{R}, \mathbb{R}^q) | R(\frac{d}{dt})w = 0\} \tag{50.2}$$

where the C^∞–assumption is made in order to avoid smoothness problems which are not germane to the questions posed in this note. Call Σ_R *autonomous* if $w_1, w_2 \in \mathcal{B}_R$ and $w_1(t) = w_2(t)$ for $t < 0$ implies $w_1 = w_2$. Call Σ_R *asymptotically stable* if $w \in \mathcal{B}_R$ implies $w(t) \longrightarrow 0$.

It is easy to prove that Σ_R is autonomous iff rank $(R) = q$ and asymptotically stable iff rank $(R) = q$, and ($\lambda \in \mathbb{C}$ and rank$(R(\lambda)) < q$) $\Rightarrow$ ($Re(\lambda) < 0$). Polynomial matrices that satisfy these conditions are called *Hurwitz.*

In [1] a Lyapunov theory has been developed for linear systems Σ_R. The associated quadratic Lyapunov functions are specified by 2–variable polynomial matrices $\Phi \in \mathbb{R}^{q\times q}[\zeta, \eta]$ that are symmetric, i.e. with $\Phi(\zeta, \eta) = \Phi^T(\eta, \zeta)$. Thus

$$\Phi(\zeta, \eta) = \sum_{k,\ell} \Phi_{k\ell}\zeta^k \eta^\ell \tag{50.3}$$

leads to the quadratic differential form

$$Q_\Phi(w) = \sum_{k,\ell} (\frac{d^k}{dt^k}w)^T \Phi_{k\ell}(\frac{d^\ell}{dt^\ell}w) \tag{50.4}$$

The derivative of Q_Φ is also a quadratic differential form, namely

$$\frac{d}{dt}Q_\Phi(w) = Q_{\overset{\bullet}{\Phi}}(w) \tag{50.5}$$

where $\overset{\bullet}{\Phi}(\zeta, \eta) := (\zeta + \eta)\Phi(\zeta, \eta)$. Call Φ (or better Q_Φ, but we will use Φ instead) $\mathcal{B}_R$*–nonnegative* if $w \in \mathcal{B}_R$ implies $Q_\Phi(w) \geq 0$, and $\mathcal{B}_R$*–positive* if $w \in \mathcal{B}$ implies ($Q_\Phi(w) \geq 0$, and $Q_\Phi(w) = 0$ only if $w = 0$).

In [1] it is shown that Φ is $\mathcal{B}_R$–nonnegative iff there exist $F \in \mathbb{R}^{\bullet\times q}[\zeta, \eta]$ and $C \in \mathbb{R}^{\bullet\times q}[\xi]$ such that

$$\Phi(\zeta, \eta) = F^T(\eta, \zeta)R(\eta) + R^T(\zeta)F(\zeta, \eta) + C^T(\zeta)C(\eta) \tag{50.6}$$

and $\mathcal{B}_R$–positive iff in addition (R, C) is observable (in the sense defined in [2]).

In [1] we have proven that Σ_R is asymptotically stable iff there exists a symmetric $\mathcal{B}_R$–nonnegative $\Phi \in \mathbb{R}^{q\times q}[\zeta, \eta]$ with an $\mathcal{B}_R$–negative derivative $\overset{\bullet}{\Phi}$. Further, if Σ_R is asymptotically stable, then for all symmetric $\Psi \in \mathbb{R}^{q\times q}[\zeta, \eta]$ there exists a $\Phi \in \mathbb{R}^{q\times q}[\zeta, \eta]$ such that $\overset{\bullet}{\Phi} \overset{\mathcal{B}_R}{=} \Psi$ where $\overset{\mathcal{B}_R}{=}$ signifies that $(w \in \mathcal{B}_R) \rightarrow (Q_{\overset{\bullet}{\Phi}}(w) = Q_\Psi(w))$. Moreover, Ψ $\mathcal{B}_R$–negative implies that Φ is $\mathcal{B}_R$–non-positive. Finally, linear algorithms are developed that allow to compute Φ from Ψ and R.

3 Nonlinear systems

The problem posed in this note consists of generalizing these results to a special class of nonlinear systems, those described by polynomial differential equations. The study of such systems in control theory has been initiated by Sontag in [6] and further developed by many authors. Let $f : (\mathbb{R}^q)^{n+1} \to \mathbb{R}^\bullet$ be a polynomial map. View it as the differential polynomial

$$f(w, \dot{w}, \ddot{w}, \cdots) \tag{50.7}$$

that specifies the system of differential equations

$$f(w, \frac{d}{dt}w, \cdots, \frac{d^n}{dt^n}w) = 0 \tag{50.8}$$

In the language of [2], (50.8) defines the dynamical system

$$\Sigma_f := (\mathbb{R}, \mathbb{R}^q, \mathcal{B}_f) \tag{50.9}$$

with behavior

$$\mathcal{B}_f := \{w : \mathbb{R} \to \mathbb{R}^q | \text{ (50.8) holds}\} \tag{50.10}$$

Because of the polynomial nature of f, we call (50.9) a *differential algebraic system*.

The problem posed is to examine what part of the linear theory can be generalized to such nonlinear differential-algebraic systems. The well developed algorithms for differential polynomial algebra (as exemplified by the theory of Gröbner bases [3]) make it feasible that important progress is feasible in this area, following the line of research initiated by Michel Fliess [4, 5] and others. See also [7].

In particular, the following questions are put forward:

- Under what conditions is Σ_f autonomous (or a suitably adapted version of this notion)?
- Let Φ be a polynomial in $w, \dot{w}, \ddot{w}, \cdots$. When is Φ $\mathcal{B}_f$–nonnegative? $\mathcal{B}_f$–positive? In other words, what is the analogue of (50.6)?
- Develop Lyapunov theory and algorithms for the construction of Lyapunov functions for differential algebraic systems.

4 References

[1] J.C. Willems and H.L. Trentelman, On quadratic differential forms, *SIAM Journal on Control and Optimization*, to appear.

[2] J.C. Willems, Paradigms and puzzles in the theory of dynamical systems, *IEEE Transactions on Automatic Control*, volume 36, pages 259-294, 1991.

[3] D. Cox, J. Little, and D. O'Shea, *Ideals, Varieties and Algorithms: An Introduction to Computational Algebraic Geometry and Commutative Algebra*, 2nd edition, Springer Verlag, 1997.

[4] M. Fliess, Automatique et corps différentiels, *Forum Mathematicae*, pages 227-238, 1989.

[5] M. Fliess and S.T. Glad, An algebraic approach to linear and nonlinear control, pages 223-267 of *Essays on Control: Perspectives in the Theory and Its Applications*, edited by H.L. Trentelman and J.C. Willems, Birkhäuser, 1993.

[6] E.D. Sontag, *Polynomial Response Maps*, Springer-Verlag, 1979.

[7] Y. Wang and E.D. Sontag, Algebraic differential equations and rational control systems, *SIAM Journal on Control and Optimization*, volume 30, pages 1126-1149, 1992.

51

Performance lower bound for a sampled-data signal reconstruction

Yutaka Yamamoto* and Shinji Hara**

*Department of Applied Analysis and Complex Dynamical Systems
Graduate School of Informatics
Kyoto University
Kyoto 606-8501
JAPAN
yy@kuamp.kyoto-u.ac.jp

**Department of Computational Intelligence and Systems Science
Tokyo Institute of Technology
4259, Nagatsuta-cho, Midori-ku
Yokohama 226
JAPAN
hara@cs.dis.titech.ac.jp

1 Description of the problem

Consider the block diagram Fig. 1.

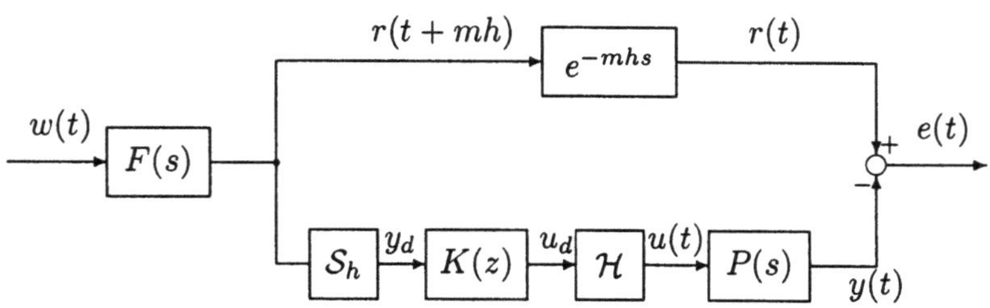

FIGURE 1. Signal Reconstruction Problem

This diagram represents an optimal signal reconstruction problem in digital signal processing. The analog input signal first goes through the anti-aliasing prefilter $F(s)$, and then gets sampled to become a discrete-time signal y_d with sampling rate h. It is then processed by a digital filter $K(z)$ and converted back to an analog signal via the hold device $\mathcal{H}$ and amplifier $P(s)$. In signal processing, reconstruction with some amount of delay is often tolerable, so that the output of $P(s)$ is compared with the delayed analog signal $e^{-mhs}F(s)w$. For simplicity, the length of the delay is assumed to be an integer multiple of the sampling period h.

An H^∞ type signal reconstruction problem is the following: Let $\mathcal{T}_{ew}(K)$ denote the (lifted) transfer function operator (see, e.g., [2]) from w to e with controller $K(z)$. The objective is to find, for a given $\gamma > 0$, a discrete-time controller $K(z)$ such that $\|\mathcal{T}_{ew}(K)\|_\infty < \gamma$.

Our problem is the following:

> Find the performance lower bound γ_{min} for a given delay length m. This γ_{min} is a function of m. How small can this γ_{min} be made by taking m large? In other words, what is the behavior of this lower bound as $m \to \infty$?

2 Motivation and Related Results

The generalized plant associated with this sampled-data H^∞ norm minimization problem is

$$\left[\begin{array}{c|c} e^{-mhs}F(s) & -P(s) \\ \hline F(s) & 0 \end{array}\right].$$

Due to the delay operator, this problem is infinite-dimensional, but one can reduce it to a finite-dimensional design problem for a non-causal controller $z^m K(z)$ with generalized plant

$$\left[\begin{array}{c|c} F(s) & -P(s) \\ \hline F(s) & 0 \end{array}\right].$$

Let (A, B, C) be a minimal realization of $F(s)$, and let $\mathcal{D}$ denote the convolution operator

$$\mathcal{D} : L^2[0,h] \to L^2[0,h] : w \mapsto \int_0^\theta Ce^{A(\theta-\tau)}Bw(\tau)d\tau.$$

It is shown in [4] that the above problem can be reduced to a norm-equivalent discrete-time (not a sampled-data) problem if the given performance bound $\gamma > \|\mathcal{D}\|$. This restriction is necessary for invoking a standard technique in sampled-data theory. The crucial step is to guarantee the positivity of the operator $\gamma^2 I - \mathcal{D}^*\mathcal{D}$.

In the usual sampled-data context (when $m = 0$), this restriction is almost necessary since $\mathcal{D}$ represents the direct feedthrough term of the plant which cannot be reduced by sampled feedback (because it is at least one step delayed). However, in the present situation, we allow a delay in signal reconstruction, and this means that the controller can react after it observes what had happened. In other words, we have a perfect preview (of

m sampling period), and it is plausible that the sensitivity can be reduced further below $\|\mathcal{D}\|$.
This intuition may be supported by the following facts:

- In the purely discrete-time case where $F(s)$, $P(s)$ are replaced by discrete-time transfer functions and e^{-mhs} by z^{-m}, the performance lower bound approaches 0 as $m \to \infty$. This is shown, for example, in [1]. See also [3] for results in ℓ^p, $p = 1, 2, \infty$ norm for the discrete-time case.

- In the purely continuous-time case where we replace the sampled controller by a continuous-time controller, the norm can also be reduced.

On the other hand, due to the limitation of the capability of sample/hold device, it is hardly expected that the performance lower bound should approach zero even when we let $m \to \infty$.
The same problem appears in a preview tracking by a feedforward sampled-data controller $K(z)$ for a continuous-time stable plant $P(s)$, where the future reference command $r(t + mh)$ instead of the current reference command $r(t)$ is available to reduce the tracking error $e(t) = r(t) - y(t)$ in Fig. 1. $F(s)$ corresponds to the reference generator. Note that if we allow two-degree-of-freedom sampled-data controllers, then the preview control problem for unstable plants can also be reduced to the same problem for stable plants. This fact can be readily seen by considering the parameterization of all stabilizing sampled-data controllers for $P(s)$.

3 References

[1] T. Chen and B. A. Francis, "Design of multirate filter banks by $\mathcal{H}_\infty$ optimization," *IEEE Trans. Signal Processing*, SP-43, pp. 2822–2830 (1995)

[2] T. Chen and B. A. Francis, *Optimal Sampled-Data Control Systems*, Springer, 1995.

[3] M. E. Halpern, "Preview tracking for discrete-time SISO systems," *IEEE Trans. Autom. Control*, AC-39, pp. 589–592 (1994)

[4] P. P. Khargonekar and Y. Yamamoto, "Delayed signal reconstruction using sampled-data control," *Proc. 35th IEEE CDC*, pp. 1259–1263 (1996)

52

Coprimeness of factorizations over a ring of distributions

Yutaka Yamamoto

Department of Applied Analysis and Complex Dynamical Systems
Graduate School of Informatics
Kyoto University
Kyoto 606-8501
JAPAN
yy@kuamp.kyoto-u.ac.jp

1 Description of the problem

Let $R := \mathcal{E}'(\mathbf{R}_-)$ denote the ring (algebra) of distributions having compact support in $(-\infty, 0]$, $p, q \in R$, and let $\hat{p}, \hat{q}$ denote their (two-sided) Laplace transforms. Notice that by the well known Paley-Wiener theorem [2] they are entire functions of exponential type.
We say that the pair (p, q) is *spectrally coprime* if $\hat{p}(s)$ and $\hat{q}(s)$ have no common zeros in the whole complex plane $\mathbf{C}$. It is *approximately coprime* if there exist $x_n, y_n \in R$ such that

$$p * x_n + q * y_n \to \delta \text{ in the sense of distributions.} \tag{52.1}$$

Here δ denotes the Dirac delta distribution. It is *(exactly) coprime* if

$$p * x + q * y = \delta \tag{52.2}$$

is satisfied for some $x, y \in R$. Note that in the Laplace transform domain (52.2) takes the familiar form

$$\hat{p}(s)\hat{x}(s) + \hat{q}(s)\hat{y}(s) = 1. \tag{52.3}$$

We are interested in questions of the following type:

1. Clearly exact coprimeness is stronger than approximate coprimeness, and the latter is yet stronger than spectral coprimeness. Are they strictly stronger than the subsequent ones?

2. What happens if we replace $\mathcal{E}'(\mathbf{R}_-)$ by $\mathcal{E}'(\mathbf{R})$ of distributions with compact support (not necessarily contained in $(-\infty, 0]$)?

3. If $\hat{p}$ and $\hat{q}$ belong to the ring of polynomials $\mathbf{R}[s, e^s]$, then does (52.1) imply (52.2)?

4. Fix a multiplicative subset M of $\mathcal{E}'(\mathbf{R}_-)$ and consider the ring $R = M^{-1}\mathcal{E}'(\mathbf{R}_-)$ of fractions by M. What can we say about comprimeness over this R?

2 Motivation and Related Results

Such problems arise in realization and stabilization of a certain class of distributed parameter systems.
Consider a fractional representation $q^{-1} * p$ for an impulse response function considered over R, where the inverse is taken with respect to convolution. Various systems, for example delay systems, admit representation in this ring [5]. The constraint on the support is related to causality and finite-length memory of the system. Aside from this consideration, one may also take the ring $\mathcal{E}(\mathbf{R}_+)$ of distributions having compact support in $[0, \infty)$, which is clearly isomorphic to R. However, for realization, R gives an analogy for the ring of polynomials and is more convenient.
A standard observable realization can be associated to this factorization, and this realization is approximately reachable if and only if the factorization if approximately coprime [5]. The questions above are thus related to various reachability tests. For example, spectral coprimeness corresponds to the property that every modal subsystem is reachable.
Actually, in $R = \mathcal{E}'(\mathbf{R}_-)$, these notions are known to be distinct. To see approximate coprimeness is stronger than spectral coprimeness, it is enough to take $p := \delta_{-1}$, $q := \delta_{-2}$, where δ_a denotes the delta distribution placed at a. While their Laplace transforms e^s, e^{2s} never vanish (and hence the pair being spectrally coprime), any combination $p*x+q*y$, $x, y \in R$, has support contained in $(-\infty, -1]$, so that (52.1) cannot be satisfied. This is due to the non-vanishing common factor e^s. Actually, in this class of factorization, any common factor is known to be either of common-zero type or of this type e^{as} for some $a > 0$. In fact, a pair (p, q) is approximately coprime if and only if i) (p, q) is spectrally coprime and ii) either p or q is not identically zero in any neighborhood of 0 [6].
Approximate coprimeness is again strictly weaker than exact coprimeness. To see this, take $q := \delta_{-1}$, and let p be any C^∞ function with compact support in $[-1/2, 0]$ such that p is not identically zero in any neighborhood of 0. Then by the result cited above, this pair is approximately coprime. On the other hand, this pair cannot be exactly coprime. To see this, suppose $p * x + q * y = \delta$. Observe first that the support of $\delta_{-1} * y$ is contained in

$(-\infty, -1]$. Hence only x in a neighborhood of the origin can contribute to δ. But since p is C^∞, this is impossible because the convolution of a C^∞ function with any distribution must be a C^∞ function.

This example does not work when R is replaced by $\mathcal{E}'(\mathbf{R})$. Indeed, in this case, $q = \delta_{-1}$ itself is invertible with inverse δ_1, and the Bezout identity $p * 0 + q * \delta_1 = \delta$ trivially holds. This gap is related to the causality constraint, and coprimeness in this case seems to be related to the behavioral controllability issues. The work of Rocha and Willems [1] seems closely related to this issue. However, it seems still not entirely clear whether approximate coprimeness or spectral coprimeness implies exact coprimeness in this larger ring. This will give us a more complete picture for behavioral controllability.

The arguments above should clarify the meaning of our third question. For delay systems with commensurable delays, the transfer functions admit fractional representations over $\mathbf{R}[s, e^s]$ (assuming that the unit delay length is 1). Surely in this case, the counterexample above does not work. In all known examples for delay systems, approximate coprimeness seems to imply exact coprimeness, and this is the conjecture here. We do not require that the solutions x, y belong to $\mathbf{R}[s, e^s]$. They can be just elements of $\mathcal{E}'(\mathbf{R})$.

Note that this is not a simple issue settled by Hilbert's Nullstellensatz. Write z for e^s, and consider $p = z-2$, $q = sz$. As elements in the polynomial ring $\mathbf{R}[s, z]$, (p, q) is not coprime because $s = 0, z = 2$ gives a common zero. However, once we impose the constraint $z = e^s$, this is easily seen to be impossible. Thus the question here is rather analytical than being algebraic.

The final question is related to stabilization. If M gives a "stable" subset, then $M^{-1}\mathcal{E}'(\mathbf{R}_-)$ gives a ring of stable transfer functions in the present setting [4]. The question then asks what kind of condition assures the existence of a stabilizing controller constructed in this ring. For related questions, see [3]. We also remark that the present situation is entirely different from that for the ring H^∞. If it were H^∞, the well-known corona condition

$$\inf_{\Re s \leq 0} \min\{|\hat{p}(s)|, |\hat{q}(s)|\} > 0.$$

gives a condition for exact coprimeness. However, this is not applicable to the present situation.

3 References

[1] P. Rocha and J. C. Willems, "Behavioral controllability of delay-differential systems," *SIAM J. Control and Optimiz.*, 35, pp. 254–264 (1997)

[2] L. Schwartz, *Theorie des Distributions*, Hermann, Paris (1966)

[3] M. Vidyasagar, H. Schneider and B. A. Francis, "Algebraic and topological aspects of feedback stabilization," *IEEE Trans. Autom. Control*, AC-27, pp. 880–894 (1982)

[4] M. Vidyasagar, *Control System Synthesis: A Factorization Approach*, MIT Press, Cambridge, MA (1985)

[5] Y. Yamamoto, "Pseudo-rational input/output maps and their realizations: a fractional representation approach to infinite-dimensional systems," *SIAM J. Control and Optimiz.*, 26, pp. 1415-1430 (1988)

[6] Y. Yamamoto, "Reachability of a class of infinite-dimensional linear systems: an external approach with applications to general neutral systems," *SIAM J. Control and Optimiz.*, 27, pp. 217-234 (1989)

53

Where are the zeros located?

Hans Zwart[1]

Faculty of Mathematical Sciences
P.O. Box 217
7500 AE Enschede
THE NETHERLANDS
twhans@math.utwente.nl

1 Description of the problem

The transfer function of a single-input, single-output, linear system is defined as the quotient (for zero initial conditions) between the Laplace transform of the output signal and the Laplace transform of the input signal. For finite-dimensional systems the transfer function is rational, whereas for infinite-dimensional system it is a non-rational function. Examples of infinite-dimensional systems are e.g. systems described by partial differential equations or systems described by delay differential equations.
For the transfer functions of single-input, single-output, linear infinite-dimensional systems we want to have information about the location of the zeros without calculating them explicitly. For instance, we want to obtain a criteria that tell us, on bases of the system data, that all the zeros are in the left-half plane. Note that for rational transfer functions there are the well-known Routh-Hurwitz conditions.

2 Problem motivation

In adaptive control for infinite-dimensional systems it is almost standard to assume that the system is minimum phase, meaning that all its zeros are in the open left half plane, see Logemann and Townley [4]. This assumption is needed in order to have that high-gain output feedback will stabilize the system. Furthermore, since the zeros are in direct correspondence to the one-dimensional controlled-invariant subspaces contained in the kernel of

[1]This paper was written while the author visited MFO. This visit was supported by the Volkwagenstiftung (RiP program at Oberwolfach)

the output operator, knowledge about the zeros gives information about the controlled-invariant subspaces. For instance, based on the location of the zeros, Zwart [5] gave necessary and sufficient conditions for the existence of the largest controlled invariant subspace contained in the kernel of the output operator.
For the poles of a transfer function, corresponding to a partial differential equation, or delay-differential equations, or a linear state-space system, many techniques can be used to gain information on their location. For instance, if the system is conservative with respect to some norm, then there will be no poles in the right-half plane. At the moment one does not have similar results for the zeros.

3 Available results and desired extensions

At the moment there is very little known about the zeros of a non-rational transfer function. On the one hand, one has many results from complex analysis, but they do not take into account that the function is the transfer function of a system. On the other hand, only a few results are known that relate the zeros to the underlying system, see Kobayashi et al. [2, 3], and Zwart & Hof [6]. Below we list their main results. Before proceeding we introduce some notation. By $\Sigma(A, B, C)$ we denote the state linear system described by the abstract differential equation on the state space Z

$$\begin{aligned} \dot{z}(t) &= Az(t) + Bu(t) \\ y(t) &= Cz(t). \end{aligned}$$

We assume that A is the infinitesimal generator of a C_0-semigroup on Z, and that B and C are bounded operators.
A transfer function will be denote by G. For the system $\Sigma(A, B, C)$ we have that $G(s) = C(sI - A)^{-1}B$. For background on infinite-dimensional state linear system, we refer to Curtain and Zwart [1].

1. If the transfer function G is positive real, i.e., $G(s) + \overline{G(s)} > 0$ for s in the right-half plane, then it is (trivially) clear that it has no zeros there. A class of positive real transfer functions is given by state linear systems of the form $\Sigma(A, B, B^*)$, where $A^* + A \leq 0$. Here A^* denotes the adjoint operator of A.

2. If $G(s) = B^*(sI - A)^{-1}B$, where A is self-adjoint, then the zeros are real and the largest zero is less then the largest pole. Furthermore, if G has the additional property that it can be written as an infinite sum of first order systems, i.e., $G(s) = \sum_n \frac{r_n}{(s-\lambda_n)}$, then between every pair of poles there is exactly one zero. Note that the special form of the transfer function, $G(s) = B^*(sI - A)^{-1}B$, implies that all r_n's are positive.

3. Assume that G has the infinite sum expansion as given in item 2 with $r_n > 0$ and λ_n real. If for some $p \geq 1$ we have that $\sum_n |\lambda_n|^p r_n < \infty$, then
$$\sum_n \frac{(\lambda_n - \mu_n)^2 |\lambda_n|^{p-2}}{r_{n+1}} < \infty,$$
where μ_n is the zero between λ_{n+1} and λ_n. Hence one has extra information concerning the (asymptotic) location of the zeros.

As interesting open problems, one has the following:

1. Do similar results as listed in 1. and 2. hold for systems that do not have the output operator equal to B^*, or that have less structure on the operator A?

   Since the results as listed above are stated via the transfer function, it is very easy to prove these results for state linear systems $\Sigma(A, B, C)$ for which the transfer function G can be written as $G(s) = B_1^*(sI - A_1)^{-1}B_1$, for some A_1 and B_1 with A_1 satisfying the properties of 1. or 2.

2. Does a result as listed in 3. hold for a larger class of $\{r_n\}$? That is, does the above mentioned result also hold if a positive sequence $\{r_n\}$ does not satisfy $\sum_n |\lambda_n|^p r_n < \infty$ for some $p \geq 1$.

   This question is important if one studies the location of the zeros for system in which B and/or C is unbounded.

3. Are there easy checkable conditions on the system data, i.e., A, B, and C, that imply that the zeros of its corresponding transfer function are in the left-half plane?

4 References

[1] R.F. Curtain and H. Zwart, *An Introduction to Infinite-Dimensional Systems Theory*, Text in Applied Mathematics, Vol. 21, Springer Verlag, New York, 1995.

[2] T. Kobayashi, "Zeros and Design of Control Systems for Infinite-Dimensional Systems", *International Journal of System Science*, 23, pp. 1507–1515 (1992).

[3] T. Kobayashi, J. Fan and K. Sato, "Invariant Zeros of Distributed Parameter Systems", *Transaction of the Institute of Systems, Control and Information Engineers*, 8, pp. 106–114 (1995).

[4] H. Logemann and S. Townley, "Adaptive control of infinite-dimensional systems without parameter estimation: An overview", *IMA journal of mathematical control and information*, 14, pp. 175-206 (1997).

[5] H.J. Zwart, *Geometric Theory for Infinite-Dimensional Systems*, Lecture Notes in Control and Information Sciences, 115, Springer Verlag, Berlin (1989).

[6] H. Zwart and M. Hof; "Zeros of infinite-dimensional systems", *IMA Journal of Mathematical Control and Information*, 14, pp. 85-94 (1997).

Index of Authors